MODELS IN ECOLOGY

MODELS IN ECOLOGY

J. MAYNARD SMITH
Professor of Biology, University of Sussex

CAMBRIDGE UNIVERSITY PRESS

CAMBRIDGE
LONDON   NEW YORK   NEW ROCHELLE
MELBOURNE   SYDNEY

Published by the Press Syndicate of the University of Cambridge
The Pitt Building, Trumpington Street, Cambridge CB2 1RP
32 East 57th Street, New York, NY 10022, USA
296 Beaconsfield Parade, Middle Park, Melbourne 3206, Australia

© Cambridge University Press 1974

ISBN 0 521 20262 0 hard covers
ISBN 0 521 29440 1 paperback

First published 1974
Reprinted 1975
First paperback edition 1978
Reprinted 1979

Printed in Great Britain at the
University Press, Cambridge

TO SHEILA

CONTENTS

[vii]

PREFACE

Ecology is still a branch of science in which it is usually better to rely on the judgement of an experienced practitioner than on the predictions of a theorist. Theory has never played the role that it has in population genetics, perhaps because there is nothing in ecology comparable to Mendel's laws in genetics. Nevertheless there has in recent years been an increase both in the volume and the relevance of theoretical work in ecology, an increase which owes much to the influence of Robert MacArthur. This book has been written in the twin convictions that ecology will not come of age until it has a sound theoretical basis, and that we have a long way to go before that happy state of affairs is reached.

The book was conceived during a six months' visit to my laboratory by Dr Don Landenberger of the University of California. Our plan was to include models of two kinds; mathematical models of a 'strategic' kind aimed at an understanding of the general properties of ecosystems, and laboratory models designed with the same aim in view. We hoped that the mathematical and experimental models would illuminate one another. Unfortunately Don Landenberger decided after his return to California that other pressures and demands on his time made it impossible for him to be a joint author. His decision has undoubtedly weakened the book, particularly on the experimental side. In the event, I decided to go on with the project on my own, but I am well aware that I should never have been able to undertake it if it had not been for the stimulus of discussions with him during his visit.

The book is aimed at anyone with a serious interest in ecology. Although there is a good deal of mathematics, I have concentrated on making clear the assumptions behind the mathematics and the conclusions to be drawn, and have as far as possible omitted mathematical proofs and derivations. I hope therefore that the book will be comprehensible to anyone with a minimal familiarity with mathematical notation. A number of the mathematical models have not been published elsewhere: the main new contributions are the analysis of delays due to development time in Chapter 3B, the models of migration in Chapter 6, the discussion of the effects on the stability of ecosystems of competition at different trophic levels in Chapter 10C, the analysis of character displacement and the evolution of specialists and generalists in Chapter 11B, and the models of territorial behaviour in Chapter 12.

As I have already indicated, my major debt is to Dr Don Landenberger. Professor G. N. Ward showed me how to handle delay difference equations. Dr C. Strobeck has read through the manuscript and has detected a number of errors. I have also greatly benefited from discussions with Dr R. C. Lewontin, Dr R. M. May, Dr M. Slatkin, Mr D. T. Streeter, Dr C. Strobeck and Dr J. D. Thomas. To them, and to the many other colleagues who have put up with my conversation during the past year, I shall always be grateful.

Sussex University   J. Maynard Smith
August 1973

1 INTRODUCTION

A. Models in ecology

Mathematical descriptions of ecological systems may be made for two quite different purposes, one practical and the other theoretical. Descriptions with a practical purpose I will call 'simulations'. If, for example, one wished to know how many fur seals can be culled annually from a population without threatening its future survival, it would be necessary to have a description of that population, in its particular environment, which includes as much relevant detail as possible. At a minimum, one would require age-specific birth and death rates, and knowledge of how these rates varied with the density of the population, and with other features of the environment likely to alter in the future. Such information could be built into a simulation of the population, which could be used to predict the effects of particular management policies.

The value of such simulations is obvious, but their utility lies mainly in analysing particular cases. A theory of ecology must make statements about ecosystems as a whole, as well as about particular species at particular times, and it must make statements which are true for many different species and not just for one. Any actual ecosystem contains far too many species, which interact in far too many ways, for simulation to be a practicable approach. The better a simulation is for its own purposes, by the inclusion of all relevant details, the more difficult it is to generalise its conclusions to other species. For the discovery of general ideas in ecology, therefore, different kinds of mathematical description, which may be called models, are called for. Whereas a good simulation should include as much detail as possible, a good model should include as little as possible. A model cannot be used to predict the future behaviour of whole ecosystems, or of any of the species composing it. Perhaps the best way of explaining what kinds of questions can be answered by the use of models is to list some of the questions I attempt to answer in this book. An an example, in Chapter 6 I attempt to answer the question, if in any one place a predator–prey interaction leads to large amplitude fluctuations and to extinction, is it possible that the two species can coexist in a large region? This leads to the further questions, what patterns of interaction and of relative mobility are most likely to lead to stability? As a second example, in later chapters I attempt to answer the question, is an ecosystem more likely to persist if each species interacts directly with a large number of others, or only with a few?

These examples will indicate the kind of question to which I think models are relevant. One general point about these questions – or rather, about the kind of answers we hope for – is worth making. As Levins (1968) has pointed out, when studying complex systems we should not look for assertions which are true of all systems or of all species. Instead we should look for the causes of differences of behaviour between different species or systems.

To answer this kind of question, we need models which are as simple as possible. In all actual ecosystems, migration, territorial behaviour, climatic fluctuations and intraspecific genetic variation are all relevant. But we will not for this reason put all these features into all our models. Instead we adopt the method of the experimental scientist, which is to vary one factor at a time, and to do so in a system which is otherwise as simple as possible. To take a particular example, when considering the effects of many as opposed to few species interactions on the stability of an ecosystem, it is clearly necessary to consider a model ecosystem with many species. But the model should be as simple as possible in other respects, for two reasons; first, if it is too complicated it is difficult to discover the behaviour of the model, and second, it is important when comparing the behaviour of two model ecosystems to be sure that it is the multiplicity of species interactions, and not some other factor, which is the relevant difference.

So far, I have discussed only mathematical models. Sometimes the behaviour of such models can be discovered analytically; more often it is necessary to resort to computer simulation. It should be clear that a resort to a computer does not convert a model into a simulation in the sense used above.

Ecologists have also made extensive use of 'biological models'; that is to say, of laboratory ecosystems composed of actual organisms, such as *Paramecium*, *Tribolium*, *Daphnia* and *Drosophila*. Such biological models form a useful bridge between mathematical models and real ecosystems. Their main value has been not so much in checking the deductions made from mathematical models, but in suggesting phenomena that the mathematician should be able to explain. It will be obvious to the reader of this book that the mathematical models of, for example, migration were suggested by Huffaker's experiments (1958) on mites, and not the other way round. Mathematical and biological models complement one another; in the absence of biological models, the mathematical treatment would tend to become more and more abstract and general, and to that extent more difficult to apply. In the absence of mathematical treatment, it would be difficult to see the general relevance of particular biological models. Because of this complementary nature, I have not in this book written different sections on the two classes of model, but instead have grouped together the biological and mathematical models relevant to a particular problem.

B.   Some possible approaches to theoretical ecology

In analysing any complex system, the crucial decision lies in the choice of relevant variables. In this book, I have chosen to concentrate on the numbers of individuals of different species, and to develop a dynamics on this basis. This is the approach pioneered by Volterra (1926), and followed by most ecologists since. It is important to remember that it is not the only possible approach. Lotka (1925), in addition to working with species numbers, also considered as possible variables the energy in different trophic levels of an ecosystem, and also the distribution of particular chemical substances (*e.g.* fixed nitrogen). This approach has also been followed up, and we have in particular a considerable body of data on 'ecological energetics', giving not only the quantities of energy locked up at any moment in plants, herbivores, decomposers and so on, but also rates of flow between these levels. As yet, these data do not constitute a 'dynamics', which would require equations giving rates of flow as functions of other variables. But the development of a dynamics of energy flow in ecosystems, and of other chemical 'cycles', is clearly on the way, and will be an important contribution to ecological theory.

A third possible approach is to take as the relevant variables the frequencies of genes within species; that is, to choose the same variables as are relevant to evolution theory. At first sight this may seem a council of despair, since it introduces additional complexity into a system already too complex for us to analyse. This would, however, be a misunderstanding. The logic behind introducing genetic variables into ecology is as follows. The properties of species (*e.g.* their patterns of reproduction or of migration) are not chosen at random, but are those which have evolved under the influence of natural selection. Thus evolution theory places some restraints on the properties which it is either plausible or necessary to consider in an ecological theory.

In practice, it may be possible to introduce such evolutionary restraints into ecological theory without giving a complete formulation in terms of gene frequencies and selection coefficients, because it is often reasonable to assume that if some particular phenotype would be favoured by selection, then that phenotype will be established. For example, Lack (1954) argued that a pair of birds will produce that number of eggs which maximises the number of their surviving progeny; most geneticists would accept this argument without demanding evidence that there are gene differences which affect egg numbers. More sophisticated arguments of the same kind are those concerning niche separation (MacArthur and Wilson, 1967, pp. 167 *et seq.*), and the relative advantages of genetic and physiological methods of adapting to fluctuating environments (Lewontin, 1961; Levins, 1968). In this book, I adopt the genetic approach when considering territorial behaviour (Chapter 12), and the evolution of ecosystems (Chapter 11).

C. Two-species interactions, or complexity *per se*?

Having chosen species numbers (or rather, species densities) as the relevant variables, a choice still remains between two possible strategies. The first is to concentrate on the interactions between a pair of species (or at the most a small number of species – say up to four), while assuming that everything apart from that pair (*i.e.* the densities of other species, the climate, etc.) either remains constant or fluctuates randomly. Such a pair is said to be 'isolated'. One can then investigate the effects on the behaviour of this isolated species pair of such factors as age structure, territorial behaviour, migration, etc. The alternative is to consider from the outset an ecosystem composed of many interacting species.

I adopt the first of these strategies in Chapters 2 to 6, and the second in Chapters 7 to 10. Which approach is the more relevant cannot be prejudged, since it depends on the answer to the following question. Does the extent to which actual ecosystems show properties of persistence or stability depend on the fact that the pairwise interactions between species would likewise, in isolation, lead to stability and persistence? The alternative view, expressed for example by Elton (1958), is that the behaviour of the whole system depends essentially on its complexity, and the behaviour of individual species depends on the whole. No final answer can be given to this question, although my present opinion is that the former view is more nearly true than the latter.

D. A classification of species interactions

It is convenient to classify the direct interaction between a pair of species into three categories, as follows:

(*a*) Competition (−, −). Each species has an inhibiting effect on the growth of the other.

(*b*) Commensalism (+, +). Each species has an accelerating effect on the growth of the other.

(*c*) Predation (+, −). One species, the 'predator', has an inhibiting effect on the growth of the other, the 'prey'; the prey has an accelerating effect on the predator.

According to these definitions, the host–parasite and the plant–herbivore interactions would be classified as 'predation'. This has the advantage of logic and simplicity. It has the disadvantage, which I have not avoided, that in thinking of examples of the $(+, -)$ interaction one is led to think of examples of actual predation rather than of parasitism or of herbivore–plant interaction. This is particularly unfortunate because, until very recently, most biologists were trained either as botanists or as zoologists, and are therefore in any case reluctant to give the plant–herbivore interaction the attention it deserves.

An 'inhibiting' effect should be understood to mean either an increase in the death rate or a decrease in the birth rate; the same is true in reverse of an 'accelerating' effect.

In this book, I have paid considerably more attention to predation than to competition or commensalism, particularly when considering single-species interactions. The reason for this does not lie in the relative importance of competition and predation, but only in the fact that the dynamics of predator and prey are more difficult to understand than of a pair of competitors.

E.   Types of stability

In the main, this book is concerned with the stability of complex ecosystems. I here introduce and define some of the terms, which I shall use in this discussion. A fuller account is given by Lewontin (1969).

The main mathematical idea I shall use is that the behaviour of a system can be represented by a set of trajectories in n-dimensional space. Those unfamiliar with this idea will probably find it easier to grasp if they first read Chapter 2B, where it is applied to a particular problem. In essence, the state of a system at any instant is defined by the values of a number of variables $x, y, z, \ldots$ These variables will usually be the densities of the species composing the system. The state of the system can then be represented as a point in n-dimensional phase space, where n is equal to the number of variables. To each point in this space we can attach a vector, or arrow, indicating how the system will move. These vectors can be joined to form trajectories which show the long-term behaviour of the system.

For this to work, there must be a unique vector associated with each point in the space; otherwise the behaviour is indeterminate. This means that we must choose variables which are sufficient to determine the behaviour. Thus suppose we were interested in the movement of a weight constrained to move along a straight line, and attached to a spring. We could specify the position of the weight at any instant by a single variable, x, giving its distance from a fixed point. But a knowledge of x is not sufficient to tell us how x will change in the immediate future; to know that we must also know the velocity v. A knowledge of x and v is sufficient to determine the future value of both x and v; we do not also need to know the acceleration, since knowing x, and the properties of the spring and the weight, we can calculate it. We can then represent the future behaviour of the weight as a trajectory in the two-dimensional (x, v) space. Thus x and v together are sufficient to determine the behaviour, but x by itself is not. As I point out in Chapter 3 the densities $x, y, \ldots$ of species composing an eco-system are usually not sufficient to determine its behaviour, although they are so in the simple models considered in Chapter 2.

A *stationary point* or *equilibrium point* is one which has associated with it a vector of zero length. That is, if a system is at a stationary point at one instant in time, it will remain at that point at the next instant.

A stationary point may be *stable* or *unstable*. If, when a system is disturbed slightly from a stationary point, it returns to that point, then the stationary point is said to be stable; more precisely, it possesses '*neighbourhood stability*'. If, on the other hand, when a system is disturbed slightly from a stationary point, it continues to move further away from that point, that point is unstable.

Suppose that no matter where a system starts from (provided none of the variables is zero) it moves to a particular stationary point, then that point has *global* stability. Figure 1 shows an unstable stationary point B, and a stationary point A which has neighbourhood, but not global, stability. The region of phase space from which the system will move to A is called the *region of*

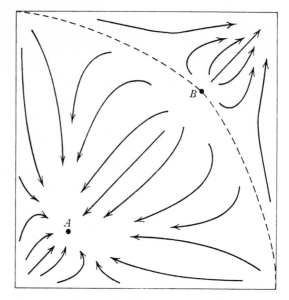

Figure 1. A system with a stable (A) and unstable (B) equilibrium point. The particular behaviour shown describes the frequences p, p' of a gene in two niches with habitat selection (Maynard Smith, 1966).

attraction of A; this introduces the obvious analogy of a landscape in which hollows represent stable points and peaks unstable ones, and the behaviour of the system is represented by a rolling ball.

Mathematically, the easiest aspect of stability to investigate is neighbourhood stability. By considering only small displacements from an equilibrium point, it is possible to linearise the equations and so find their behaviour. This behaviour will follow one of four patterns – divergent oscillation, convergent oscillation, convergent exponential and divergent exponential (figure 2). Thus consider the simple recurrence relation

$$X_{n+1} = X_E + k(X_n - X_E). \tag{1}$$

This gives the value X_{n+1} of a variable in the $(n+1)$th generation in terms of its value X_n in the nth generation, and of the constants X_E and k. The equilibrium value is clearly $X_n = X_E$. If we write $X_n = X_E + x_n$, equation (1) becomes

$$x_{n+1} = kx_n, \tag{2}$$

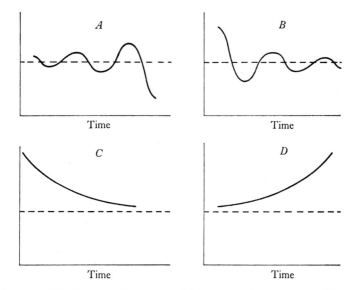

Figure 2. The four possible types of behaviour close to an equilibrium point: A, divergent oscillation; B, convergent oscillation; C, convergent exponential; D, divergent exponential. In each diagram, the full line represents a variable whose equilibrium value is the broken line.

where x_n measures the difference between the actual value of X and its equilibrium value. Then clearly:

if $k < -1$, there is a divergent oscillation (DO),
if $-1 < k < 0$, there is a convergent oscillation (CO),
if $0 < k < +1$, there is a convergent exponential (CE), and
if $k > +1$, there is a divergent exponential (DE).

In other words, as we change k continuously, the behaviour of the system undergoes the transformations DO $\rightarrow$ CO $\rightarrow$ CE $\rightarrow$ DE. Starting in the stable region CO or CE, the changes in k required to produce a divergent oscillation are precisely the opposite of those required to produce a divergent exponential. This conclusion has been reached for a recurrence relation with a single variable, rather than for a system with n variables described by n differential equations. Nevertheless, the conclusions drawn do have some validity for complex systems. Close to a stationary point, the same four patterns of behaviour are possible. Unfortun-

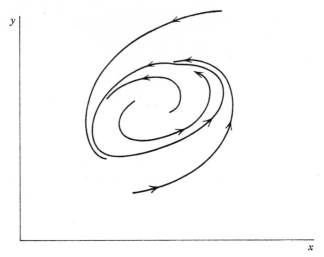

Figure 3. Limit cycle behaviour.

ately, it is no longer true that as one alters the coefficients in the equations (*i.e.* as one alters the relative importance of different interactions) one always undergoes the transformations in the sequence DO → CO → CE → DE. However, it is often the case that if one has a system with a stable stationary point, the changes one must make in that system to produce a divergent oscillation are opposite to those required to produce a divergent exponential. An example is given in figure 24. The importance of this fact will emerge particularly in Chapter 10. But one general point is worth making here: it will be found that competitive interactions tend to cause divergent exponential instability, and predator–prey interactions to cause oscillating instability.

These four categories exhaust the types of behaviour open to linear systems. But linear equations will only be an adequate description of the behaviour of a system for small displacements from an equilibrium point. More complex patterns of behaviour are displayed by non-linear systems. Of these, the most interesting is *limit cycle* behaviour. Figure 3 shows limit cycle behaviour in two-dimensional phase space. Instead of a stationary point, there is a stable cycle – *i.e.* an oscillation of stable amplitude, so that if the system starts at a point outside the limit cycle, its amplitude of oscillation declines, and if inside its amplitude increases, until the stable cycle is achieved. Thus a system may have a stable pattern of behaviour, and yet not display numerical constancy of any of the state variables.

A second reason why a dynamically stable system may not display numerical constancy is that the system is continuously perturbed from without. In that case the actual numbers are likely to fluctuate irregularly about their equilibrium values.

The terms I have mentioned so far in this section have a mathematical rather than an ecological meaning; they refer to the properties of models rather than of actual ecosystems. In conclusion, I introduce two terms which refer to observable features of actual ecosystems. An ecosystem is *taxonomically persistent* if the species composing it remain the same for long periods (*i.e.* long compared to the generation time of the species); it is *numerically persistent* if the relative numbers of individuals in different species either remain constant, or return regularly to the same ratios, for long periods. Obviously, persistence in this sense is a relative rather than an absolute property; no ecosystem persists for ever, but some are more persistent than others.

I have omitted one possible category of behaviour – a so-called conservative oscillation – *i.e.* an oscillation of constant amplitude, the amplitude depending on the initial conditions (figure 4). Although there are differential equations (*e.g.* $d^2x/dt^2 + ax = 0$) which lead to such behaviour, there are reasons why we do not expect to meet them in the real world. Suppose we alter the relevant equation by adding an arbitrary function $\phi(x)$, multiplied by a small constant ϵ, to give $d^2x/dt^2 + ax + \epsilon\phi(x) = 0$. This new equation no longer describes an oscillation of constant amplitude. The behaviour will now be either a convergent or a divergent oscillation. The dynamical system described by the original equation, $d^2x/dt^2 + ax = 0$, is said to be *structurally unstable*, because the smallest arbitrary change in the equations in any direction will alter the whole topology of the trajectories in phase space. It will be clear why we do not expect to meet such systems.

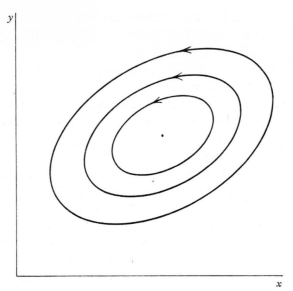

Figure 4. A conservative oscillation.

F. Stochastic and deterministic models

Most of the models considered in this book are deterministic. This needs some justification, which will now be given. Readers unfamiliar with mathematical ecology would be well advised to leave this section until after reading the rest of the book; it can then be treated as a retrospective justification.

There are two ways in which a deterministic model fails to mirror ecological reality: first, it assumes an infinite population size, and second, it ignores random fluctuations in the environment with time. These points will now be discussed.

As an example of a deterministic model in ecology, consider the equation

$$dN/dt = aN, \tag{3}$$

where N is the number of individuals at time t, and a the intrinsic rate of increase. The equation has the solution $N = N_0 e^{at}$. The equation assumes that in the short time interval δt, each individual gives rise to a fraction $a\,\delta t$ of new individuals. The corresponding stochastic model makes the more plausible assumption that in the interval δt, an individual produces one offspring with probability $a\delta t$, and no offspring with probability $1 - a\delta t$. With these assumptions, it is possible (Bartlett, 1966) to calculate the full probability distribution of offspring at time t, and hence the average population

$$\hat{N} = N_0 e^{at},$$

and also the variance of N,

$$\text{var}(N) = N_0 e^{2at}(1 - e^{-at}).$$

The variance of N measures the differences in population size between replicate populations with the same initial size N_0. Note that the mean of N corresponds to the value given by the deterministic model, and that as $t \to \infty$, the coefficient of variation of N,

$$[\text{var}(N)]^{\frac{1}{2}}/\hat{N} \to N_0^{-\frac{1}{2}}.$$

Hence if N_0 is large, the deterministic model gives an adequate picture of the behaviour of the population.

The use of deterministic rather than stochastic models can only be justified by mathematical convenience. It must always be remembered that if a deterministic model predicts that the number of one or more species in an ecosystem will from time to time fall to a low value, then the corresponding stochastic model would give a non-negligible probability of extinction. The general attitude I shall take in this book is to assume that if the deterministic model shows a stable equilibrium, the corresponding stochastic model would predict long-term survival, whereas if the deterministic model shows no equilibrium or an unstable one, the stochastic model would predict extinction with a high probability.

The second deficiency of deterministic models is that they ignore environmental randomness. The mathematical difficulties of models which allow for such randomness are formidable. I now briefly summarise the account of the problem given by May (1971a); anyone seriously interested should refer to that paper.

The simplest model, corresponding to equation (3), is

$$dN/dt = [a+y(t)]N, \tag{4}$$

where $y(t)$ is a random variable with mean zero. $y(t)$ is taken to be 'white noise'; that is, the random distribution from which the $y(t)$

are taken is the same at all times, and there is no correlation between the fluctuations at successive instants. The assumption of no serial correlation seems unrealistic, but in fact all it means is that the fluctuations are correlated only over times which are short compared to other relevant time scales in the system (for example, $1/a$).

The probability distribution of N has a mean,

$$\hat{N} = N_0 \exp{(at)},$$

which is identical to that without environmental fluctuations. The variance of N is given by

$$\mathrm{var}(N) = N_0{}^2 \exp{(2at)}[\exp{(\sigma^2 t)} - 1],$$

where σ^2 is the variance of $y(t)$. Hence

$$[\mathrm{var}(N)]^{\frac{1}{2}}/\hat{N} = [\exp{(\sigma^2 t)} - 1]^{\frac{1}{2}}.$$

Thus population fluctuations become relatively more severe as time goes on; this is a reflection of the fact that the deterministic system (3) has no steady state.

Finally, it can be shown that if $\sigma^2 > 2a$, the probability that the system will become extinct tends to unity as time tends to infinity. We therefore have the odd conclusion that if environmental fluctuations are large enough, then a population whose expected size is increasing is nevertheless certain ultimately to become extinct. The same conclusion has been reached by Lewontin and Cohen (1969) for a system described by a finite difference equation.

We now turn to single-species systems which in the deterministic form do have a stable stationary point. When environmental variation is allowed for, such a system would be described by an equation of the general form

$$dN/dt = g[N, y_1(t), y_2(t), \ldots, y_k(t), \ldots],$$

where $y_1(t), \ldots, y_k(t), \ldots$ are environmental parameters which may contain white noise. For such an equation, it will often be the case that a steady equilibrium probability distribution, independent of t, will ultimately be reached. This equilibrium probability distribution in the stochastic environment corresponds to the stable equilibrium point in a deterministic environment.

The method of calculating the equilibrium probability distribution, if it exists, is given by Kimura (1964). It has been shown by Levins (1969, 1970) that the general effect of environmental fluctuations is to lower the population density below what it would be in a deterministic environment, and if they are severe enough to cause extinction.

May (1971 a) then considers the effects of environmental noise on an ecosystem of m species, densities $N_1, \ldots, N_m$. In the deterministic form, such a system can be described by equations of the form

$$\mathrm{d}N_j/\mathrm{d}t = F_j[N_1, N_2, \ldots, N_m],$$

with $j = 1, 2, \ldots, m$.

It is shown in Chapter 10 that the behaviour of such a system depends on the $m \times m$ matrix A of the elements a_{jk} measuring the interactions between the species. If all the eigenvalues of A have negative real parts, the system is stable.

In the stochastic version of this model, it is supposed that their interaction coefficients are subject to random fluctuations, as was a in equation (4). For such a model, does a stable equilibrium probability distribution exist? It turns out that, provided the fluctuations are not too large (in a sense to be defined in a moment), the conditions for an equilibrium probability distribution are identical to the conditions for a stable equilibrium in the deterministic case; that is, the matrix A must be stable. Very roughly, this limitation requires that

$$\sigma^2 \ll \bar{a}, \tag{5}$$

where σ^2 is a measure of the variances and covariances of the fluctuations of the interaction coefficients, and $\bar{a}$ is the typical magnitude of the matrix elements a_{jk}.

This important theorem is, as May remarks, 'a substantial justification for our playing with deterministic models'.

2 PREDATOR–PREY SYSTEMS WITHOUT AGE STRUCTURE

A. Exponential and logistic growth

In this chapter I shall discuss mathematical models of population growth and of predator–prey interaction which are based on a number of simplifying assumptions, as follows:

(i) The density of a species – that is, the number of individuals per unit area – can be adequately represented by a single variable. This ignores age differences, and also differences of sex and genotype.

(ii) Changes in density can adequately be described by deterministic equations.

(iii) The effects of interactions within and between species are instantaneous. In the predator–prey interaction, this means that the delay between the moment a predator eats a prey, and the moment when the ingested material is converted into part of a new predator, is ignored.

If these assumptions are made, it is possible to describe predator–prey interactions by ordinary differential equations, about which a great deal is known. The effects of relaxing assumptions (i) and (iii) are considered at some length in Chapters 3 and 4. Assumption (ii) was discussed at the end of the last chapter. The main justification for deterministic equations is that they are easier to handle than stochastic ones. But it must be remembered that if the deterministic equations predict that the population density will reach a steady state, an actual finite system is likely to fluctuate irregularly about that state; if the deterministic equations predict large amplitude oscillations, an actual finite population may well go extinct at the low points of the oscillation.

The simplest differential equation describing population growth is

$$dx/dt = rx, \tag{6}$$

where x is the population density at time t, and r is a constant. It has the solution $x = x_0 e^{rt}$, where x_0 is the density at time

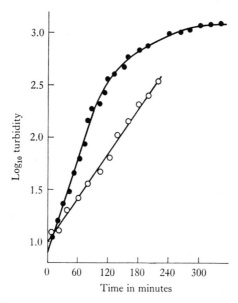

Figure 5. Growth of two cultures of *E. coli*, closed circles, in nutrient broth; open circles, in synthetic medium. Turbidity is proportional to number of cells per unit area; a turbidity of 100 units ($\log_{10}$ turbidity $= 2$) corresponds to a bacterial density of approximately 10^8 cells/ml. Exponential growth in nutrient broth ceases after approximately 60 minutes. From Stent (1963).

$t = 0$. This is a good description of the growth of a bacterial colony before the medium is exhausted (figure 5).

The implication of equation (6) is that if at any time a large sample of individuals was observed during a short time interval δt, the fraction which reproduced during that interval would be equal to $r\delta t$, where r is constant. This in turn implies that the 'age distribution' – that is, for bacteria, the fraction of individuals at different stages of the division cycle – remains constant with time. This need not necessarily be so. For example, if the cells are synchronised, equation (6) will not be an adequate description. However, it was shown by Lotka (1925) that a reproducing population will approach a stable age distribution provided that the age-specific birth and death rates (for bacteria, the distribution of inter-division times) remains constant. This in turn is likely to be

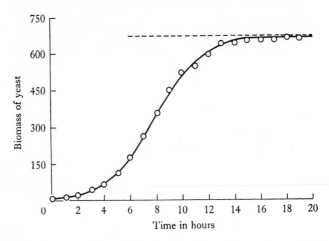

Figure 6. A comparison of the growth of yeast in a culture with logistic growth, from Allee *et al.* (1949).

the case provided that environmental conditions are constant, resources are present in excess, and evolutionary changes do not occur.

Equation (6) can hold true only for a limited period of time; ultimately an increasing population will exhaust its resources. The population may settle down to some steady value; it may fluctuate, regularly or irregularly; or it may decline. An equation often used to describe the former behaviour, settling to a steady value, is the logistic equation

$$\mathrm{d}x/\mathrm{d}t = ax - bx^2, \tag{7a}$$

or
$$\mathrm{d}x/\mathrm{d}t = rx(1 - x/k). \tag{7b}$$

The justification for this equation is that it is the simplest differential equation with the two required features:

(i) when x is small, the equation reduces to (6), and growth is exponential, and

(ii) as t increases, x approaches a steady value without oscillations.

Figure 6 shows a comparison between the growth of yeast cells in culture and growth predicted by the logistic equation.

In equation ($7b$), r is referred to as the intrinsic rate of increase and k as the carrying capacity. In this chapter I will use 'carrying capacity' mainly to refer to the equilibrium density attained by a prey species in the absence of predation; there is an obvious interest in the ratio between the carrying capacity so defined, and the equilibrium density of the prey in the presence of predation.

The logistic equation is best regarded as a purely descriptive equation. Thus equation (6) for exponential growth, although it does adequately describe the growth of some populations, can also be thought of as being deduced from a knowledge of the behaviour of the individuals forming the population, although such a derivation depends on the assumption (or proof) that there is a stable age distribution. In contrast, equation (7) was not derived from any knowledge of or assumptions about the precise way in which the reproduction of individuals is influenced by density; it is merely the simplest mathematical expression for a particular pattern of growth. In ecology as in other sciences, it is important to distinguish between equations, or 'laws', whose justification is that they describe the observed relation between two or more variables, and those which have in addition some 'microscopic' justification in terms of the known or postulated behaviour of the components of the system.

B. Volterra's equations

Volterra (1926) considered the following equations describing the interactions between a prey species, density x, and its predator, density y:

$$\left.\begin{aligned} \dot{x} &= ax - bx^2 - cxy, \\ \dot{y} &= -ey + c'xy \end{aligned}\right\} \tag{8}$$

The assumptions lying behind these equations, in addition to those considered at the beginning of this chapter, are as follows:

(i) In the absence of predation, the prey species would follow the logistic equation, with intrinsic rate of increase a and carrying capacity a/b.

(ii) The rate at which prey are eaten is proportional to the product of the densities of predator and prey.

Later in this chapter I discuss alternative assumptions about the rate of predation, and some of the observational evidence. Volterra's assumption would be true if one or both species move at random; if, when they meet, there is a constant probability that the predator will kill the prey; and if the time spent by the predator in consuming the prey is negligible. The assumption is similar to that made in chemical kinetics for the rate at which two different molecules will react to form one.

How will a system described by these equations behave? At any time t, the state of the system is fully described by the values of x and y; to each state there corresponds a point in the (x, y) plane, which is called the 'phase plane' (or 'phase space', since a system may require more than two variables for its description). If to each point in the phase plane we can attach an arrow, indicating the direction in which a system at that point will move, then we can join up these arrows to form trajectories which will tell us how the system will move.

The first step is to plot in the phase plane the lines for which $\dot{x} = 0$ (and hence arrows are parallel to the y axis) and $\dot{y} = 0$ (arrows parallel to the x axis). Thus

$$\dot{x} = 0 \quad \text{when} \quad a - bx - cy = 0,$$
$$\dot{y} = 0 \quad \text{when} \quad -e + c'x = 0.$$

The first question to ask is whether a stationary state exists, with $\dot{x}$ and $\dot{y}$ simultaneously zero. This requires

$$x = e/c'; \quad y = \frac{a}{c} - \frac{be}{cc'}.$$

Here, x is necessarily positive, but y is positive only if $a/c > be/cc'$, or $a/b > e/c'$; this inequality states that for there to be a steady state with prey and predator both present, the carrying capacity of the prey, a/b, must be high enough to support the predator.

To find the dynamics of the system, arrows must be inserted as in figure 7. To do this we note that for $x > e/c'$, $\dot{y}$ is positive, and for $x < e/c'$, $\dot{y}$ is negative. Also, for points in the plane above the line $a - bx - cy = 0$, $\dot{x}$ is negative, and below it $\dot{x}$ is positive. This enables us to insert the arrows shown in figure 7A. In figure 7B

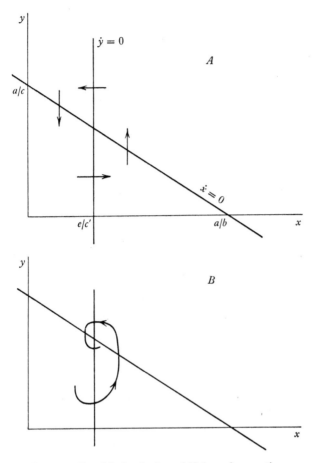

Figure 7. Graphical solution of Volterra's equations;
for explanation see text.

these arrows have been joined up to form a trajectory, which
is a spiral converging towards the stationary point. The convergent
nature of the spiral follows from the angle at which the $\dot{x} = 0$ and
$\dot{y} = 0$ lines intersect. This point is illustrated in figure 8. This
rather loose argument, based on geometrical intuition, can be
justified analytically for the equations considered in this book.
The meaning of a convergent anticlockwise spiral is illustrated in
figure 9; both predator and prey numbers oscillate with decreasing

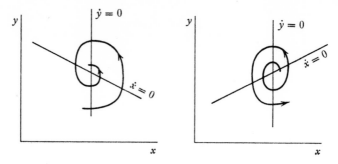

Figure 8. Relation between the stability of an equilibrium and the slope
of the line $\dot{x} = 0$.

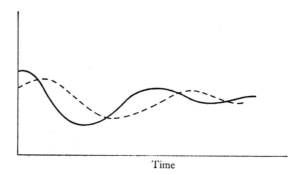

Figure 9. Solution of Volterra's equations with self-limitation of the
prey: full line, density of prey; broken line, density of predator.

amplitude, the predator oscillations lagging in phase behind the
prey.

Suppose now that the prey species is limited only by the predator; thus $b = 0$, and the prey increases exponentially in the
absence of the predator. The behaviour of such a system is shown
in figure 10. The oscillations are of constant amplitude, depending
on the initial conditions; a system started close to its steady state
will have small amplitude oscillations, and one started far from its
steady state will have large amplitude oscillations. Such a system
is called 'conservative', because there is a quantity which is
conserved during the motion, as energy is conserved in simple
harmonic motion.

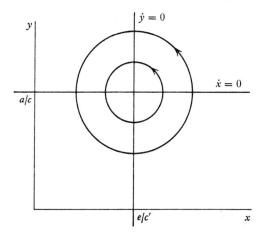

Figure 10. Solution of Volterra's equations with no self-limitation of the prey.

The term $-bx^2$ in equation (8), expressing the inhibiting effect of a species on its own growth, is referred to as a 'damping' term, by analogy with engineering usage, in which a damper is a shock-absorber or other device whose function it is to reduce the amplitude of oscillations. In ecology, the main factor reducing oscillations is the presence of such self-inhibiting effects. In equation (8) a damping term was included for the prey, but not for the predator. The logic behind this is that the predator is supposed to be limited only by its food – that is, by the prey. The possibility of other types of limitation on predators is considered in Chapter 12.

C. Leslie's equations

An alternative formulation of the predator–prey equation was suggested by Leslie (1948), as follows:

$$\left.\begin{array}{l} \dot{x} = ax - bx^2 - cxy, \\ \dot{y} = ey - fy^2/x. \end{array}\right\} \tag{9}$$

The equation for the prey is identical to Volterra's equation with damping. The equation for the predator also resembles the logistic equation, but the second term has been modified to allow

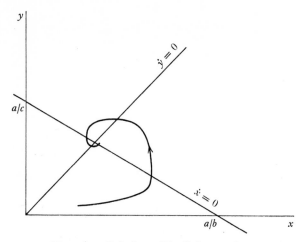

Figure 11. Solution of Leslie's equations.

for the density of the prey. Thus if x/y is large (many prey per predator) the predator increases exponentially; if $x/y = f/e$, then the predator is at equilibrium; if x/y falls below f/e, the predator decreases in numbers.

It is clear from figure 11 that the equations lead to rapidly damped oscillations.

It is also apparent from figure 11 that the essential difference between Volterra's and Leslie's formulation of the problem is as follows: for Volterra, whether the predator increases or decreases in number depends only on the density of prey, whereas for Leslie it depends on the number of prey per predator. If it is remembered that equations (8) and (9) refer to instantaneous rates of change, not to long-term prospects, it is clear that Volterra's formulation is usually to be preferred. A second reason for preferring Volterra's equations is that they do relate the rate of increase of the predators to the rate (cxy) at which prey are being eaten, whereas in Leslie's formulation there is no relationship between the rate at which a predator eats and the rate at which it reproduces.

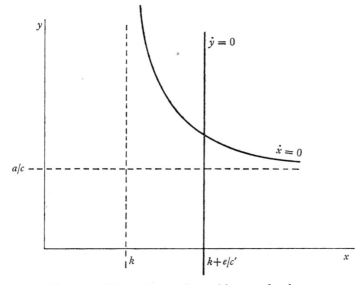

Figure 12. Volterra's equations with cover for the prey.

D. Some special cases

(i) *The effect of cover for the prey*

Suppose that some number x_r of the prey can find some cover or refuge which makes them inaccessible to the predator. Volterra's equations without damping then become:

$$\left.\begin{array}{l} \dot{x} = ax - cy(x - x_r), \\ \dot{y} = -ey + c'y(x - x_r). \end{array}\right\} \quad (10)$$

There are two cases to consider. First, the number of prey in cover may be a constant fraction of the total; that is, $x_r = kx$. This is equivalent to replacing the constants c and c' in equations (9) by $c(1-k)$ and $c'(1-k)$. Clearly this does not alter the nature of the equilibrium.

A more interesting conclusion emerges if the number of prey in cover is constant; that is, $x_r = k$. In this case, the curve of $\dot{x} = 0$ in the (x, y) plane is $y = ax/c(x-k)$, and of $\dot{y} = 0$ is $x = k + e/c'$. These curves are shown in figure 12, from which it is apparent that the effects of cover are stabilising, since it changes a conservative into a convergent oscillation.

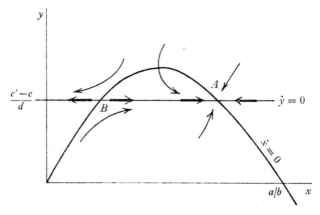

Figure 13. Graphical representation of predator–prey system when the predator has a constant food intake, and both predator and prey are self-limited.

(ii) *A predator with constant food intake*

Further insight into the predator–prey relation can be obtained by making a different assumption concerning the rate at which a predator takes prey. For a warm-blooded predator it seems reasonable to assume that an individual predator takes prey at a constant rate, the rate being determined by the food requirements of the predator and not by the density of prey. The appropriate equation for the prey then becomes

$$\dot{x} = (a-bx)x - cy. \tag{11}$$

The corresponding equation for the predator would be

$$\dot{y} = -ey + c'y = (c'-e)y. \tag{12}$$

This is an equation for exponential increase, if $c' > e$, or decrease. Thus if we assume that a predator has a constant food intake independent of prey density, then the predator cannot be limited by its prey (unless we allow for a declining conversion efficiency or increasing mortality as the prey density declines, and this we have not done). Hence if an equilibrium is to exist, we must suppose that the predator is in some way self-limited; that is, that there is a damping term in the equation for y. Thus

$$\dot{y} = (c'-e)y - dy^2. \tag{13}$$

The behaviour of the system described by equations (11) and (13) is analysed graphically in figure 13. There are two stationary points with x and y non-zero. A is a stable non-oscillatory equilibrium and B an unstable equilibrium. Whether the system will reach the stable equilibrium A depends on the initial conditions.

This model is unrealistic, since the assumption that each predator takes prey at a constant rate regardless of prey density must break down when the prey are sufficiently rare. It may nevertheless give an adequate picture in the region of the stable equilibrium point A.

A method of analysing the predator–prey interaction for more realistic patterns of predation was developed by Rosenzweig and MacArthur (1963) and by Rosenzweig (1969), and is outlined in a modified form in the next section.

E. A more general case – the Rosenzweig–MacArthur model

Consider the equations:

$$\left.\begin{array}{l} \dot{x} = f(x) - \phi(x, y), \\ \dot{y} = -ey + k\phi(x, y), \end{array}\right\} \tag{14}$$

where $f(x)$ is the rate of change of x in the absence of predators, $\phi(x, y)$ is the rate of predation, k the conversion efficiency of prey into predator, and e the mortality rate of the predator. It has been assumed that the predator is limited only by its prey, but in other respects the equations are fairly general. They can be made simpler, and easier to handle, if it is assumed that the rate at which an individual predator takes prey depends only on the prey density, and is independent of the density of predators. Equations (14) then become

$$\left.\begin{array}{l} \dot{x} = f(x) - y\phi(x), \\ \dot{y} = -ey + ky\phi(x). \end{array}\right\} \tag{15}$$

Then $\dot{y} = 0$ when $k\phi(x) = e$; that is, when x has some constant value, as in figure 14. In what circumstances will equations (15) lead to a divergent oscillation? It is clear from figure 14 that divergent oscillations will arise if

(i) the predator is an efficient hunter, so that it can survive at

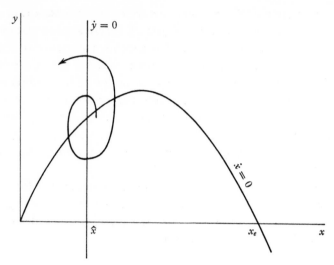

Figure 14. Divergent predator–prey oscillation.

a prey density $\hat{x}$ which is a small fraction of the carrying capacity x_e, and

(ii) if the curve $\dot{x} = 0$ in the (x, y) plane has a maximum, or 'hump' as Rosenzweig has called it.

To obtain the $\dot{x} = 0$ curve we proceed as in figure 15. First we plot $f(x)$, and a set of curves $y\phi(x)$ for varying values of y, against x. The intersection of the curves of $f(x)$ and $y_i\phi(x)$ gives the values of x for which $\dot{x} = 0$ when $y = y_i$. Hence we construct the curve $\dot{x} = 0$ in the (x, y) plane.

Clearly, there is a maximum if and only if some of the $y\phi(x)$ curves intersect the $f(x)$ curve twice. This can happen for one of two reasons:

(1) The $f(x)$ curve has the form shown in figure 16. To understand what this means, we recall that $\dot{x} = f(x)$ gives the rate of growth of a species in the absence of predation. We can write this as $\dot{x} = Rx$, where $R = f(x)/x$ is the reproductive rate of the prey. Note that a curve of the type shown in figure 16 implies that R declines for small x as well as for large x; this might happen because of the difficulty of finding mates when the population is sparse, or for other reasons. The conclusion is that if the prey

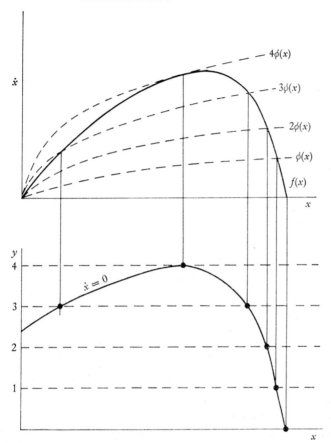

Figure 15. Construction of the $\dot{x} = 0$ line; for explanation see text.

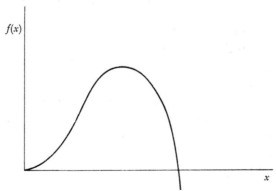

Figure 16. Productivity curve for prey species, if fertility declines at low as well as at high densities.

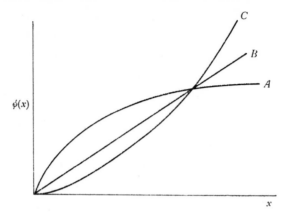

Figure 17. Possible shapes for the predator curve.

species has a reproductive rate which declines at low densities, and if it is maintained at low density by a predator, then the predator–prey interaction will lead to divergent oscillations. This is hardly surprising.

(2) It is possible for there to be a hump in the $\dot{x} = 0$ curve even if R decreases continuously with an increase of x, provided that the predator curve is of type A, figure 17, rather than B (as assumed in Volterra's equations) or C. As we shall see in the next section, a curve of type A is quite likely.

The conclusions which can be drawn from this analysis of continuously reproducing systems can be summarised briefly:

(i) The predator–prey interaction, in which the predator is limited by the supply of prey, may lead to regular fluctuations in numbers.

(ii) If the prey is resource-limited, and not limited by the predator, this will tend to damp out the oscillations.

(iii) If the predator is limited by some factor other than the prey, this will tend to damp out the oscillations.

(iv) If there is cover available which makes a constant number of prey unavailable to the predators, this will tend to damp out the oscillations.

(v) Fluctuations are likely to increase in amplitude, and perhaps lead to the extinction of one or both species, if the predator is able

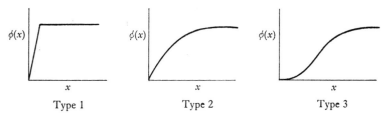

Figure 18. Three types of functional response of a predator to prey density, according to Holling (1965).

to maintain itself when the prey density is well below the carrying capacity, and if the predation curve is of type A, figure 17, rather than of types B or C.

F. The functional response of predators

The analysis in the last section has shown that the stability of a predator–prey system can depend on the shape of what Holling (1965 and earlier) has called the 'functional response' of predators to prey density; that is, of the curve of $\phi(x)$ against x, measuring the rate at which prey are taken by a predator, as a function of prey density x. Holling has classified the functional response into three types, as in figure 18. He suggests that type 1 responses are unusual, except for the special case of filter-feeding crustacea feeding on algal cells. Type 2 responses are typical of invertebrate predators. The obvious explanation for such a response is that it takes the predator a certain amount of time to kill and eat each prey. If allowance is made for a constant 'handling time' for each prey captured, one obtains an expression of the form

$$\phi(x) = ax/(1+bx), \qquad (16)$$

which fits the experimental results well.

Holling found a type 3 response for a vertebrate predator, the deermouse, *Peromyscus leucopus*, feeding on sawfly pupae in the laboratory (figure 19). It is important to appreciate that the deermice had an alternative 'prey' in the form of dog biscuits present in excess at all times. The total food intake was in fact constant; only the rate of intake of sawfly pupae varied in a type 3

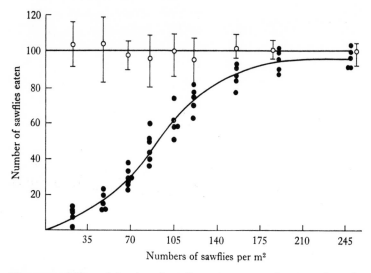

Figure 19. Effect of density of sawfly pupae on number eaten by a deer-mouse, *Peromyscus leucopus*, from Holling (1965). Closed circles, sawflies eaten; open circles, alternate food + sawflies eaten.

manner. The explanation for the S-shaped curve is that when pupae were scarce the deermice did not learn that they were present, or that they were palatable. Holling found that to make a model which predicted this type of response he had to make four assumptions:

(i) the predator has a generalised responsiveness to unfamiliar stimuli;

(ii) the rate of attack increases if the stimulus is associated with palatable prey, and decreases if it is associated with inedible prey;

(iii) learning is selective, distinguishing between stimuli;

(iv) learned associations decay if not reinforced.

It seems likely that an S-shaped functional response will be typical for vertebrate predators with alternative prey available; Holling gives some field evidence in favour of this conclusion.

The relevance of these results for the stability of predator–prey systems is as follows. In the frequent cases in which type 2 responses are found, instability of the type illustrated in figure 14 can result if the predator reduces the prey well below the carrying capacity. The relevance of type 3 responses is less clear. Such responses require that alternative prey be present, and are there-fore relevant only when a predator depends on more than one prey species. The stability of three species systems with a habit-forming predator and two prey is not easy to analyse, if only because of the difficulty of choosing plausible assumptions, and nothing will be said about such systems here. In Chapter 4C, analysing a warm-blooded vertebrate predator and its single prey species, I shall assume that a predator takes prey at a constant rate, independent of prey density, unless it is unable to find sufficient prey, in which case it starves; this assumption is to some extent supported by the constant total food intake shown in figure 19, and is in any case reasonable for a homiotherm.

G. *Paramecium–Didinium* experiments

The importance of the ratio between the equilibrium density $\hat{x}$ in the presence of predators and the carrying capacity x_e is well illustrated by Luckinbill's (1973) experiments on the ciliates *Paramecium* and *Didinium*, of which the latter is a predator on the former. Some of his results are shown in figure 20. Typically, there is first a rapid increase of the prey, followed by an increase in the predators, which capture all the prey and then starve. Prolonged coexistence (figure 20*A*) was achieved by adding methyl cellulose to the medium; this renders the medium viscous and slows down the swimming of both species. However, there was still an oscillation of increasing amplitude, ending in the extinction of the predator. Persistent coexistence (figure 20*B*) was achieved by adding methyl cellulose and at the same time halving the con-centration of food for the prey species.

The reasons for the oscillatory nature of this equilibrium cannot be understood without taking into account delayed effects, which are discussed in the next chapter. In particular, when a *Didinium* has been starved for some hours, it does not immediately resume growth and division if it then captures a prey. It is, however, possible to understand why the addition of methyl cellulose and the impoverishment of the medium should reduce the amplitude of the oscillations. Consider first the effects of methyl cellulose.

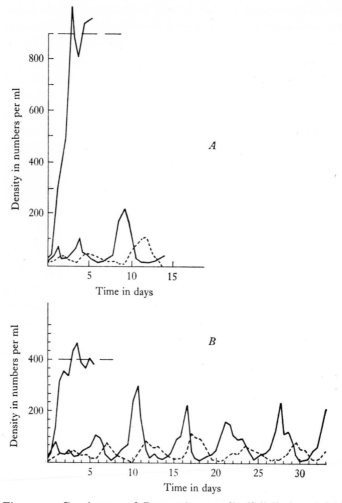

Figure 20. Coexistence of *Paramecium aurelia* (full line) and *Didinium nasutum* (broken line), after Luckinbill (1973). *A*, medium with methyl cellulose; *B*, medium with methyl cellulose and reduced food for prey. In each graph, the upper full line is for *P. aurelia* on its own.

In equation (15) we interpret y not as the number of *Didinium* but as their biomass, and e as a rate at which biomass is lost by an individual in the absence of prey. The addition of methyl cellulose will increase e relative to the rate of predation $\phi(x)$.

Now the equilibrium value $\hat{x}$ is given by the solution of the equation $k\phi(\hat{x}) = e$. An increase in e will lead to an increase in $\hat{x}$, and hence will reduce the ratio $x_e/\hat{x}$. Similarly, a reduction in the concentration of food for the prey reduces x_e (this was confirmed experimentally; the levels are shown in figure 20) and so reduces $x_e/\hat{x}$. As shown in section E, this will tend to stabilise the equilibrium.

Notice that in order to apply this (or indeed any other) mathematical model, it is not necessary that the model be a precise description of the biological situation. What in fact we do is to ask what kind of *change* in the behaviour of the model is made by a particular change in parameters, and predict that a comparable change will occur in the behaviour of the biological system.

3 BREEDING SEASONS AND AGE STRUCTURE

A. Delayed feedback

Engineers are familiar with the idea that if a system is controlled by a feedback loop in which there is a substantial delay, oscillations of large amplitude are likely to result. The same connection between oscillations and delayed regulation is apparent in everyday life. A man learning to ride a bicycle goes through a period in which he wobbles from side to side because, although he applies the appropriate corrections, he applies them too late and continues them too long. It is possible to scald and freeze oneself alternately under a shower because there is a delay between turning a tap and the corresponding stream of water reaching the skin. To give a more serious example, the booms and slumps of economic systems arise in part because of the delay between the moment when demand for a commodity exceeds supply, and the moment when some manufacturer can produce the factory space and machinery needed to meet that demand.

It is possible to be somewhat more explicit about the effects of delays on the dynamics of a system. In general, if the duration of a delay in a feedback loop is longer than the natural period of the system, large amplitude oscillations will result. The 'natural period' is to be understood as follows: if the growth of a species in the absence of regulation obeys the equation $dx/dt = rx$, then the natural period is $1/r$.

Delayed regulation in ecosystems is likely to arise from one of three causes:

(i) *Development time.* A change in the environment – for example an increase in resources – may produce an immediate change in the rate at which adults produce offspring. But it will not produce a change in the number of adults for a time T equal to the time taken for an egg to develop into an adult. If x represents the adult population, then an equation of the form

$$dx/dt = f(x)$$

[36]

should be replaced by the equation

$$\mathrm{d}x/\mathrm{d}t = f(x_{t-T}),$$

where x_{t-T} is the adult breeding population at a time T in the past.

(ii) *Discrete breeding seasons.* A species may breed only at a specific time – usually at a particular time of year. Even if individuals can survive to breed in a number of successive seasons, as is usual for mammals and birds and is true by definition for perennial plants, breeding seasons do introduce some delay in the regulative processes. If a species lives for a number of years and produces relatively few young each year, the delay time of one year due to discrete breeding seasons is likely to be short compared to the natural period of the species, and so any oscillations caused by the delay will be convergent. But if adults breeding in one season rarely or never survive to breed in the next – as in annual plants, many insects, some small mammals and birds – this has an important effect on their dynamics. Thus the equation

$$\mathrm{d}x/\mathrm{d}t = f(x)$$

should be replaced by $\quad x_{n+1} = \phi(x_n),$

where x_n is the population size in year n. It is often the case that, even though the differential equation has a stable equilibrium, the corresponding finite difference equation leads to divergent oscillations.

(iii) *Delayed response by limiting factors.* Even if a species responds instantaneously to its immediate circumstances, its numbers may nevertheless oscillate if there is a delay in the factors limiting the species. This will usually be the case if the limiting factor is itself a species subject to delays due to prolonged development or discrete breeding seasons. This can happen whether the limiting species is a predator or a resource.

Each of these three types of delay will be considered in this chapter.

The first ecologist to consider the importance of delays for the dynamics of ecosystems was Hutchinson (1948). He considered the equation $\quad \mathrm{d}x/\mathrm{d}t = x(a - bx_{t-T}).$ $\qquad$ (17)

The logic behind this equation is as follows: as the density of a species increases, the resources available to it decrease, giving rise to the logistic equation (7). However, in a real ecosystem, resources will be self-renewing, so that the actual level of resources available at any time will depend on the density of the regulated species at a time T in the past, where T is the 'development time' of the resource species. Thus the usual instantaneous logistic equation should be replaced by (17). If T is large compared to $1/a$, equation (17) leads to divergent oscillations, although the logistic equation with $T = 0$ always gives a stable non-oscillatory equilibrium.

In this chapter, however, I will first illustrate the occurrence of oscillations due to delayed regulation in a simpler system, for which both mathematical and biological models are available.

B. Delays due to development time

Nicholson (1954) showed that single-species laboratory populations of the blowfly, *Lucilia cuprina*, supplied with food at a constant rate, may nevertheless oscillate with large amplitude and regular period. Figure 21 shows fluctuations in the number of adult and larval *Lucilia*, in a cage in which unlimited food was supplied to the larvae, and a constant but limited amount of food was supplied each day to the adults. It is illuminating to consider a simple mathematical model of this experiment.

Let X be the number of adult *Lucilia* at time t, and $w\delta t$ the food supplied to the adults during the time interval δt; then w is a constant. In the interval δt, the food available per adult is $w\delta t/X$. I assume that some constant rate of food supply is required to maintain each adult; let this be m. Assuming that adult females convert excess food into eggs with constant efficiency, the eggs laid per female are $k(w/X - m)\delta t$, and the total number of eggs laid by the population, assuming a sex ratio of unity, is

$$\tfrac{1}{2}kX\left(\frac{w}{X} - m\right)\,\delta t = \tfrac{1}{2}k(w - mX)\delta t.$$

Note that if we ignore m, the food needed for maintenance, then the total rate of egg-laying is independent of the population size X.

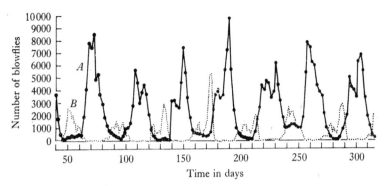

Figure 21. Numbers of blowfly *Lucilia* in a population cage. Larvae received unlimited food, and adults received a limited supply of 0.5 g of liver daily. Full line, adult population; broken line, eggs laid per day. After Nicholson (1954).

I now assume that there is a constant force of mortality C on the adults; this ignores senescence, and also the possibility that adult mortality may increase when X is large and food per adult small. The adults dying per unit time are then $CX\delta t$.

Finally, we assume a constant probability s that an egg will survive to become an adult; this is reasonable if there is unlimited food for larvae.

Then if τ is the time taken for an egg to develop into an adult,

$$\delta X = -CX\delta t + \tfrac{1}{2}ks(w - mX_{t-\tau})\delta t,$$

or $$\mathrm{d}X/\mathrm{d}t = \tfrac{1}{2}ksw - CX - \tfrac{1}{2}mksX_{t-\tau}. \tag{18}$$

Let $\hat{X}$ be the equilibrium density of X; that is to say, if X has the value $\hat{X}$ for a period τ, it will not subsequently change. Then

$$\hat{X} = \tfrac{1}{2}ksw/(C + \tfrac{1}{2}mks).$$

Let $X = \hat{X}(1 + x)$; then

$$\hat{X}\,\mathrm{d}x/\mathrm{d}t = \tfrac{1}{2}ksw - C\hat{X}(1 + x) - \tfrac{1}{2}mks\hat{X}(1 + x_{t-\tau}),$$

or $$\mathrm{d}x/\mathrm{d}t = -Cx - \tfrac{1}{2}mksx_{t-\tau}. \tag{19}$$

Clearly, if $m = 0$, $\mathrm{d}x/\mathrm{d}t = -Cx$, so that any departure from the equilibrium decreases exponentially.

In equation (19) C and m refer to the mortality and maintenance per unit time – for example, for one day. We can rewrite the equation in terms of a unit time equal to the development time τ:

$$\mathrm{d}x/\mathrm{d}t = -C\tau x - \tfrac{1}{2}mks\tau x_{t-1}, \qquad (20)$$

where τ = development time in days,

C = mortality rate per day,

$\tfrac{1}{2}mks$ = food required to maintain a fly for one day, measured in 'surviving female eggs not laid'.

Putting $C\tau = a$ and $\tfrac{1}{2}mks\tau = b$, the equation becomes

$$\mathrm{d}x/\mathrm{d}t = -ax - bx_{t-1}, \qquad (21)$$

where a and b are positive constants.

This equation is investigated analytically in the appendix to this chapter. The nature of the solution, as a function of a and b, is shown in figure 22. Since a and b are necessarily positive, we are concerned only with the positive quadrant. The effect of increasing the time delay τ is to increase a and b proportionally, and so is equivalent to moving away from the origin as shown in the figure. Thus an increase in τ has the effect of altering the behaviour of the system from a stable non-oscillatory one to a convergent oscillation, and then (provided that m/C is not too small) to a divergent oscillation. We conclude that divergent oscillations require:

(i) A substantial quantity of food be used in maintaining each adult, so that as the number of adults increases the total rate of egg-laying declines.

(ii) A time delay due to development.

Note that instability occurs only if $a = C\tau > \pi/2$, or very approximately if $\tau > 1/C$. This illustrates the generalisation made on p. 36 that instabilities arise if the delay in a feedback loop is larger than the natural period of the system.

Experimentally, Nicholson also observed sustained oscillations when unlimited food was supplied to the adults and a limited quantity to the larvae. Analysis of this case along similar lines shows that oscillations require that as the number of eggs laid on a given quantity of food increases above some optimal number, the number of adults produced should decrease. Nicholson showed

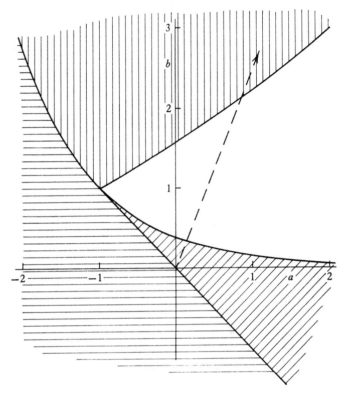

Figure 22. Stability boundaries of the equation $dx/dt = -ax - bx_{t-1}$. Horizontal hatching, divergent exponential; diagonal hatching, convergent exponential; unhatched, convergent oscillation; vertical hatching, divergent oscillation.

that this is in fact the case. When many eggs are laid, food is wasted because it is eaten by larvae which will die before pupation; this wasted food plays the same role as the maintenance food m.

These observations led Nicholson to distinguish two types of competition for limited resources:

(i) A 'scramble', as observed in *Lucilia*, in which each individual gets a share of the resource, so that, if there are more individuals than can be supported by the available resource, much is wasted, and the number surviving on a given quantity of resource declines, sometimes to zero.

(ii) A 'contest', whereby some individuals get adequate recources to enable them to survive and reproduce and others get none. Hierarchical and territorial behaviour can have this effect.

The preceding analysis confirms Nicholson's view that a scramble tends to cause oscillations, and a contest a stable equilibrium.

C. Matrix representation of population growth with age-dependent birth and death rates

A population may fall naturally into a series of discrete age classes; this is so for example for a species with an annual breeding season. Even if this is not so, it is often convenient arbitrarily to divide the population into age classes; demographic and actuarial calculations on man usually start by grouping the population into year classes. In such cases, the matrix representation suggested by Leslie (1954, 1948) is appropriate. At the time $t = 0$ the population can be represented by the column vector

$$\begin{pmatrix} n_{00} \\ n_{10} \\ n_{20} \\ \cdot \\ \cdot \\ \cdot \\ n_{m0} \end{pmatrix}.$$

Thus n_{i0} is the number of females of age i at time 0. The population structure at time $t = 1$ is then given by the equation

$$\begin{pmatrix} n_{01} \\ n_{11} \\ n_{21} \\ \cdot \\ n_{m1} \end{pmatrix} = \begin{pmatrix} F_0 & F_1 & \cdots & F_{m-1} & F_m \\ P_0 & 0 & \cdots & 0 & 0 \\ 0 & P_1 & \cdots & 0 & 0 \\ & & \cdots & & \\ 0 & 0 & \cdots & P_{m-1} & 0 \end{pmatrix} \begin{pmatrix} n_{00} \\ n_{10} \\ n_{20} \\ \cdot \\ n_{m0} \end{pmatrix}. \qquad (22)$$

In this equation $F_0, F_1, \ldots, F_m$ are the fecundities of females of different ages; thus F_i is the number of daughters born to a female of age i which survive to the next time interval, and so contribute to the number n_{01}. Similarly, P_i is the probability that a female aged i will survive to age $i+1$, so that $n_{i+1,1} = P_i n_{i0}$.

This matrix representation is convenient both for practical calculation and for analytical studies. It is not discussed further here because it has been fully treated in recent books by Pielou (1969) and Williamson (1972).

Appendix

The form and stability of the solutions of the equation

$$\dot{x}(t) = ax(t) + bx(t-1),$$

where a and b are real constants, and $x(t) = 0$ for $t < 0$ and $x(0) = 1$.

On taking Laplace transforms of both sides, we get

$$sX - 1 = aX + be^{-s}X.$$

Hence the solution for the given initial conditions is

$$x(t) = \frac{1}{2\pi i} \int_{c-i\infty}^{c+i\infty} \frac{e^{st}}{s - a - be^{-s}} \, ds. \tag{1}$$

The only singularities of the integrand are at the zeros of $s - a - be^{-s}$, so the solution is the sum of the residues of the integrand at these poles. To calculate the zeros, we put $s = \alpha + i\beta$, where α and β are real, which gives

$$\alpha + i\beta = a + be^{-(\alpha + i\beta)},$$

and on equating real and imaginary parts, we obtain the simultaneous equations

$$\alpha = a + be^{-\alpha} \cos \beta, \tag{2}$$

$$\beta = -be^{-\alpha} \sin \beta. \tag{3}$$

Clearly, one solution of (3) is $\beta = 0$, and then (2) becomes

$$\alpha = a + be^{-\alpha}. \tag{4}$$

When $b \geqslant 0$, this has just one real root, which is positive if $a + b > 0$ and otherwise is negative.

When $b < 0$, (4) may have two real roots or no real root. The condition that (4) has a double root is $be^{-\alpha} = -1$, in which case $\alpha = a - 1$ and hence

$$1 = -be^{1-a}. \tag{5}$$

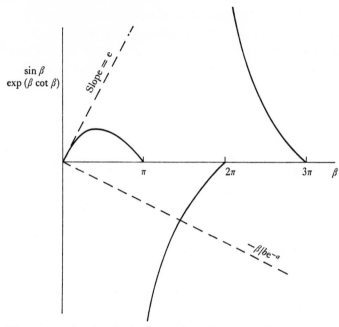

Figure 23. Graph of $\sin \beta \exp (\beta \cot \beta)$ against β. The solutions of $\beta = -b \exp (-a) \sin \beta \exp (\beta \cot \beta)$ are given by the intersections of this graph with the graph of $-\beta/b \exp (-a)$.

Therefore, when $b < 0$, there are two real roots of (4) if $b > -e^{a-1}$. The conditions for both these roots to be negative are that $a+b < 0$ and $a < 1$; otherwise one root is positive. When $b < -e^{a-1}$, there is no real root.

If α_0 is a real root of (4), then the corresponding contribution to $x(t)$ from the residue has the form $A_0 \exp (\alpha_0 t)$. Hence the conditions for exponential divergence are $a+b > 0$, or $a > 1$ and $b > -e^{a-1}$.

We now turn to the case where $\beta \neq 0$. We eliminate $b e^{-a}$ from (2) and (3) to give
$$\alpha = a - \beta \cot \beta, \tag{6}$$
and on substituting this in (3) we obtain
$$\beta = -b e^{-a} \sin \beta \exp (\beta \cot \beta), \tag{7}$$
which is an equation for β. If β is a solution of (7) then so is $-\beta$, so we may assume that $\beta > 0$ without loss of generality. Having

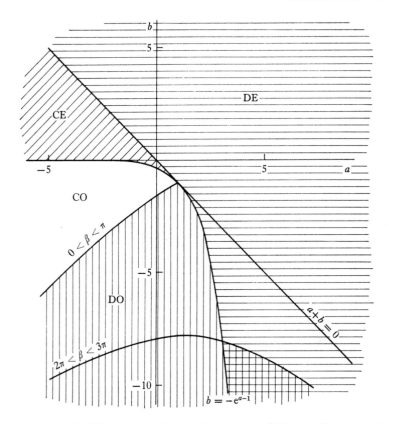

Figure 24. Stability boundaries for the equation $\dot{x}(t) = ax(t) + bx(t-1)$. Horizontal hatching, divergent exponential (DE); diagonal hatching, convergent exponential (CE); unhatched, convergent oscillation (CO); vertical hatching, divergent oscillation (DO).

found β, (6) then gives α, and the contribution to $x(t)$ from the residues at $\alpha + i\beta$ is of the form $e^{\alpha t}(A \cos \beta t + B \sin \beta t)$, and there is a contribution of this kind for each pair of poles of the type $\alpha \pm i\beta$.

The solutions of (7) can be found from the intersections of $x = \sin \beta \exp(\beta \cot \beta)$ and $x = -\beta/b e^{-a}$. From figure 23, when $b > 0$ there are solutions in the ranges of β from $\pi-2\pi$, $3\pi-4\pi$, etc. When $b < 0$ there are solutions in the ranges $2\pi-3\pi$, $4\pi-5\pi$, etc., and, if $b < -e^{a-1}$, there is also a solution in the range $0-\pi$.

The stability boundaries are the curves $\alpha = 0$ in the a, b plane, with stability when $\alpha < 0$ for all zeros of $s - a - b\,e^{-s}$, and instability otherwise. The stability boundaries for the oscillatory solutions ($\beta \neq 0$) are given parametrically by

$$a = \beta \cot \beta, \quad b = -\frac{\beta}{\sin \beta}, \quad \beta > 0.$$

The stability boundaries are shown in figure 24.

4 PREDATOR–PREY SYSTEMS WITH AGE STRUCTURE

A. Predator–prey equations with delays

(i) *Generations separate*

I consider first the case in which predator and prey species have discrete generations. If X_n, Y_n are the densities of prey and predator respectively in year n, then

$$\left.\begin{aligned} X_{n+1} &= RX_n, \\ Y_{n+1} &= rY_n, \end{aligned}\right\} \tag{23}$$

where R and r will in general be functions of X_n and Y_n. I will at first confine attention to the case in which the predator is limited only by the prey, and the prey is limited only by the predator. For example, consider the equations

$$X_{n+1} = aX_n - CX_nY_n,$$
$$Y_{n+1} = -eY_n + C'X_nY_n.$$

These are the equivalent, for discrete generations, of Volterra's equations without self-limitation. However, rather than solve this particular case, consider the general case

$$\left.\begin{aligned} X_{n+1} &= X_n \cdot G(Y_n), \\ Y_{n+1} &= Y_n \cdot H(X_n), \end{aligned}\right\} \tag{24}$$

where G and H can be any differentiable functions.

The equilibrium values $\hat{X}$ and $\hat{Y}$ can be found from the equations

$$G(\hat{Y}) = 1; \quad H(\hat{X}) = 1.$$

Let $X_n = \hat{X}(1+x_n)$; $Y_n = \hat{Y}(1+y_n)$. Then for small displacements from the equilibrium,

$$\left.\begin{aligned} x_{n+1} &= x_n + \hat{Y}\left[\frac{\partial G(Y)}{\partial Y}\right]y_n, \\ y_{n+1} &= y_n + \hat{X}\left[\frac{\partial H(X)}{\partial X}\right]x_n, \end{aligned}\right\} \tag{25}$$

where the partial differentials are evaluated at $Y = \hat{Y}$, $X = \hat{X}$.
Let $\qquad \hat{Y}[\partial G(Y)/\partial Y] = g$ and $\hat{X}[\partial H(X)/\partial X] = h$.

Then eliminating y_n from (25) gives

$$x_{n+2} - 2x_{n+1} + (1 - gh)x_n = 0. \tag{26}$$

This has the solution $\quad x_n = A\lambda_1^n + B\lambda_2^n, \tag{27}$

where λ_1 and λ_2 are the solutions of the characteristic equation
$\lambda^2 - 2\lambda + (1 - gh) = 0$; that is

$$\lambda_1 = 1 + \sqrt{(gh)}; \quad \lambda_2 = 1 - \sqrt{(gh)}. \tag{28}$$

Since this is a predator–prey interaction, g is positive and h
negative, and hence gh is negative, and so (27) describes a divergent oscillation. It follows that a predator–prey system in which
each species is limited only by the other, and in which there are
discrete breeding seasons, is necessarily unstable, whatever the
functional form of the interaction. This contrasts with the conclusion (p. 22) that with continuous reproduction the result is a
conservative oscillation.

In practice, whether a predator–prey system is stable or oscillatory will depend on a balance between damping caused by self-
limitation of the prey or predator, and delays caused by discrete
breeding seasons or development time.

(ii) *Delays caused by development time*

Wangersky and Cunningham (1957) considered the equations

$$\left.\begin{array}{l} dx/dt = ax - cxy - bx^2, \\ dy/dt = -ey + c'x_{t-\tau}y_{t-\tau}. \end{array}\right\} \tag{29}$$

These are the same as equations (8), except that it is supposed that
a time τ elapses between the moment when a prey individual is killed,
and the moment when the corresponding addition is made to the
number of adult predators. The equations were investigated by
simulation. As expected, if $b = 0$ (no self-limitation) they lead to
large amplitude oscillations. If b is not zero, the behaviour depends
on the relative magnitude of b (damping) and τ (destabilising).

More extensive simulations of predator–prey systems with
various types of delay were carried out by Caswell (1972).

B. Host–parasitoid models

One of the earliest models of a system with discrete generation describes the interactions between an insect parasitoid and its host (Nicholson and Bailey, 1935). A parasitoid is a parasite such as an ichneumon or chalcid wasp which is free-living as an adult and which lays eggs in the larvae of the host. Nicholson and Bailey considered the case in which only a single egg is laid per host larva, or, if more than one egg is laid, only one survives.

If in a particular year Y is the density of adult female parasitoids, we need to find $p(Y)$, the probability that a particular host larva will escape parasitisation. It is assumed that each female parasitoid searches an area ka ($k \geqslant 1$) and finds a proportion $1/k$ of the hosts; this is equivalent to assuming that she searches an 'effective area' a, finding all the hosts. In some area A, where $A \gg a$, there are AY parasitoid females. The probability that a particular host larva is found by one particular parasitoid is a/A, and that it escapes that parasitoid is $1 - a/A$. Provided that the parasitoids search independently (*i.e.* they are not territorial, and all host larvae are equally easy to find), we have

$$p(Y) = (1 - a/A)^{AY} = e^{-aY}. \tag{30}$$

If the number of host females is X, and the average number of eggs laid per female which survive to pupate is $2R$, then after parasitisation there are $2RX e^{-aY}$ pupae containing hosts, and $2RX(1 - e^{-aY})$ parasitised.

Assuming that there is a constant probability S that a pupa (parasitised or not) will survive the winter and give rise to an adult (host or parasite) next year, and also that both species have a sex ratio of unity, we have

$$\left. \begin{aligned} Y_{n+1} &= RSX_n[1 - \exp(-aY_n)], \\ X_{n+1} &= RSX_n \exp(-aY_n). \end{aligned} \right\} \tag{31}$$

The equilibrium values are

$$Y_E = \frac{1}{a}\log RS; \quad X_E = \frac{\log RS}{a(RS-1)}.$$

Considering small displacements from this equilibrium, it is possible to show that it is unstable and oscillatory, provided that $RS > 1$, which must be the case if the host species is to increase in numbers in the absence of the parasite.

Although large amplitude oscillations may result from host-parasitoid interactions, this is not the invariable or even the usual result. In particular, it has been shown (Hassell and Varley, 1969; Hassell, 1970) that the search area a may not be constant as suggested by Nicholson and Bailey, but may decrease with increasing parasitoid density. If so, this may have the effect of stabilizing the system. Thus from equation (30), the effective search area is given by

$$a = -\frac{1}{Y} \log p(Y). \tag{32}$$

Hassell measured the proportion of parasitisation $1-p(Y)$ of a moth, *Ephestia cantella*, by an ichneumon, *Nemeritis canescens*, in laboratory cages. Calculating the effective search area from (32), he found that it was not constant, but followed the relation

$$\log a = \log a_0 - m \log Y. \tag{33}$$

From (32) and (33) it follows that

$$p(Y) = \exp\left(-a_0 Y^{1-m}\right).$$

If we make the appropriate modifications to equations (31) and again consider small displacements from the equilibrium, we find that stability depends on the value of

$$c = \frac{(1-m)RS \log RS}{RS-1}. \tag{34}$$

If $c > 1$, there will be divergent oscillations, and if $c < 1$ there will be convergent oscillations.

If $m = 0$ (that is, no interference between parasitoids, and a constant search area) it is easy to show that $c > 1$ and hence, as stated above, the system is unstable. The critical values of m depend on RS, as follows:

RS	$1+\delta$	2	4	8	16
Critical m	0	0.28	0.53	0.58	0.66

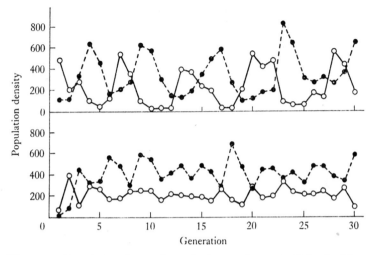

Figure 25. Population fluctuations of a host and parasitoid in the laboratory, after Utida (1957). In both cases the host species is the bean weevil,
Callosobruchus chinensis: in the top figure, the parasitoid is the braconid
wasp, *Heterospilus prosopidis*, giving large amplitude oscillations; below,
the parasitoid is another wasp, *Neocatolaccus mamezophagus*, giving small
amplitude fluctuations. Closed circles, density of host; open circles,
density of parasite.

If m is greater than the critical value, oscillations will be convergent. Measured values of m in the laboratory were in the range
0.4–0.8, so the effect can clearly be important.

The reason why a is a decreasing function of Y appears to be
that when a parasitoid comes into contact with another, it may
move away, or may temporarily stop searching, consequently
the time spent searching declines as parasite density increases.
It is not clear why we should expect the exact relationship (33).
Hassell's model may help to explain the experimental results
obtained by Utida (1957) on wasp parasites attacking the bean
weevil, *Callosobruchus chinensis* (figure 25). He was able to maintain
a cage population of the weevil with a wasp parasite for over 100
generations. There were large amplitude fluctuations, but neither
population went extinct. More unexpected, he maintained for
over 70 generations a cage population of the weevil with two species
of parasitic wasp. Coexistence of two parasitoids on a single host

would not be predicted by the Nicholson–Bailey model, but would be predicted by the Hassell–Varley model if intraspecific inter-ference of parasitoid searching is greater than interspecific interference. This raises the problem of what selective advantage intraspecific interference might confer. The only suggestion is that a parasite which does not move away from an area being searched by many others is likely to lay its eggs in hosts which have already been parasitised.

C. Warm-blooded predators

Special difficulties arise in seeing how there can be stable coexist-ence between a warm-blooded vertebrate predator and its prey. It is usually the case that both the predator and its prey have discrete breeding seasons, and this introduces delays which are likely to lead to oscillations, particularly if the total life span of one or both species is short. Unlike insect host–parasitoid systems, most of the predation occurs between breeding seasons when prey recruitment cannot replace losses, and this too will contribute to instability. Maynard Smith and Slatkin (1973) have investigated in a model system some of the factors which can stabilise such a system; this model will now be briefly described.

(i) *The prey species*

Consider a prey species which is born in one summer, which survives the winter to breed in the next summer, and which then dies. This pattern is approximately correct for many small rodents. In the nth season, a population of x_n adults alive at the beginning of the season give rise to X_n young alive at the end of the season. Then

$$X_n = Rx_n,$$

and, in the absence of mortality,

$$x_{n+1} = X_n.$$

For simplicity, mortality other than that due to predation is ignored.

The reproductive rate R is taken as

$$R = \frac{R_0}{1 + (R_0 - 1)(x_n/x_e)^c}.$$

This implies that the prey has a maximum rate of increase when rare of R_0, and an equilibrium density x_e in the absence of predation. The constant c can be chosen so as to alter the nature of the equilibrium. Thus if $F = c(R_0 - 1)/R_0$, it can be shown that for small displacements from the equilibrium:

$$F < 1: \text{no oscillations,}$$
$$1 < F < 2: \text{convergent oscillations,}$$
$$2 < F: \text{divergent oscillations.}$$

The oscillations predicted for $F > 2$ do not keep growing indefinitely, but will be sustained as a limit cycle.

(ii) *The predator species*

The predator is supposed to be limited only by the supply of prey; this limitation acts in the period between one breeding season and the next, *i.e.* during the winter.

If at the start of the nth breeding season there are y_n adult predators, and if Y_n and Z_n are the numbers of adult and young predators at the end of the season, then

$$Y_n = y_n,$$
$$Z_n = r_0 y_n,$$

where the reproductive rate of the predator, r_0, is assumed constant. The proportion of young and adult predators surviving the winter depends on the supply of prey, in a way to be discussed in the next section. At the end of the winter a constant fraction M of the surviving adult predators are assumed to die, and the surviving young predators become adult, so that

$$y_{n+1} = (1 - M) f(Y_n) + \phi(Z_n),$$

where $f(Y_n)$ and $\phi(Z_n)$ are the adult and young predators surviving the winter.

(iii) *Interaction between predators and prey*

The winter is divided into a number W of time intervals. During each interval, each predator must find and eat one prey. If it fails to find a prey it dies; if it succeeds, it does not seek for a second prey until the next interval.

If X_t, Y_t and Z_t are the numbers of prey, adult and young predators at the start of an interval, we need expressions for X_{t+1}, Y_{t+1} and Z_{t+1} at the start of the next. The searching process is represented exactly as in the Nicholson–Bailey model of host–parasitoid interaction, and leads to the equations

$$\left.\begin{aligned}
Y_{t+1} &= Y_t[\mathrm{I} - \exp(-\alpha X_t)], \\
Z_{t+1} &= Z_t[\mathrm{I} - \exp(-\beta X_t)], \\
X_{t+1} &= X_t - Y_{t+1} - Z_{t+1},
\end{aligned}\right\} \tag{35}$$

where α and β are the effective search areas of adults and young respectively. The numbers of prey and predators at the end of the winter can then be obtained by iterating these equations W times.

Finally, some allowance must be made for the number of prey eaten during the breeding season; it is assumed that this number is $S(Y_n + Z_n)$, where S is the number of prey needed to support or produce a predator during the summer.

(iv) *Behaviour of the model*

The model is not intended to represent accurately any particular predator and its prey. It has been made as simple as possible compatible with its purpose, which is to investigate factors which might influence the stability of this type of interaction.

Consider first cases in which $\alpha = \beta$; that is, in which there is no difference in hunting ability between adult and young predators. It is possible to find values of the parameters for which the two species can coexist; two examples are shown in figure 26. Figure 26*A* shows a non-oscillatory equilibrium. Note that the prey density is close to x_e; that is, the prey is limited by its resources and not by the predator. This is always so for non-oscillatory equilibria in this model; if the predator reduces the prey signifi-

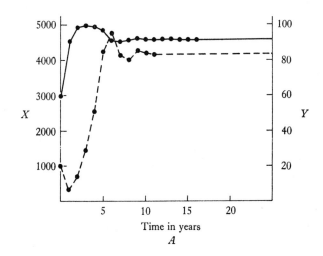

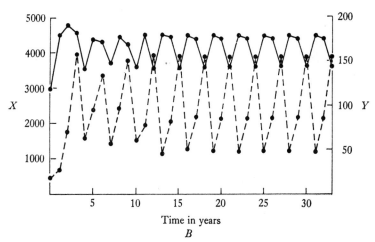

Figure 26. Two examples of persistent coexistence in a model of a verte-brate predator and its prey. In both cases $R_0 = 10$, $x_e = 5000$, $c = 1$, $r_0 = 2$, $M = 0.4$, $S = 10$ and $W = 20$. X, full line: density of prey at end of breeding season; Y, broken line: density of predator at end of breeding season. In A, the effective search area $\alpha = 0.001$; in B, $\alpha = 0.0015$. Note that a small increase in searching ability leads to oscillations; a small reduction would lead to extinction of the predator.

cantly below x_e, then large amplitude oscillations result. This conclusion is in line with but stronger than the conclusion (p. 28) drawn from the Rozenzweig–MacArthur model, and from Luckinbill's experiments (p. 35).

Although it is possible to get coexistence with $\alpha = \beta$, it requires that the search area α should lie in a rather narrow range. If α falls too low, the predator goes extinct; if α rises too high, oscillations develop which soon have a much larger amplitude than those shown in Figure 26B, and which would in a real system lead to the extinction of one or both species. The introduction of a predator may cause an intrinsically non-oscillatory prey ($F < 1$) to oscillate, but never has the opposite effect.

Since natural selection on the predator will tend to increase α, and on the prey to reduce it, it seems unlikely that α would long remain in the rather narrow range needed for persistence of the system. We have therefore to seek other stabilising effects. One change which has a large effect on the persistence of the model is to introduce a difference in hunting ability between adults and young, a difference which is known to exist in vertebrate predators which learn to hunt (e.g. Southern, 1959).

If simulations are made with $\alpha \gg \beta$, it is no longer necessary for either value to lie within such a narrow range to ensure persistence. The system may fluctuate but nevertheless persist. An example is shown in figure 27A. In effect, what happens is that in years in which the prey population at the start of the winter is low, the young predators starve early, leaving enough food for the older predators to survive the winter. This is an example of what Nicholson (1954) referred to as the difference between a 'scramble' and a 'contest'. As he argued, a 'scramble' is likely to lead to an unstable ecosystem.

It was shown (p. 25) that cover for the prey can be a stabilising factor. Unexpectedly, it does not act in this way in the present model. Thus suppose that during the winter some fixed number of prey, X_c, are exempt from attack by predators. Then in equations (35) for Y_{t+1} and Z_{t+1}, X_t must be replaced by $(X_t - X_c)$.

Simulations show that the presence of cover does not make continued coexistence more likely, and may make it less likely.

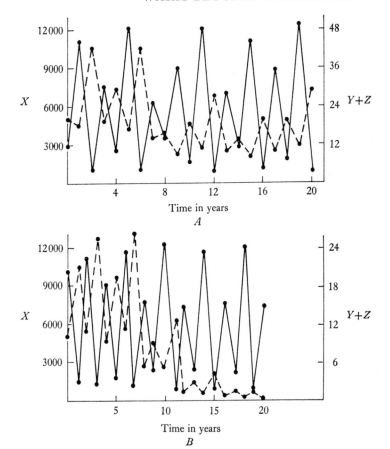

Figure 27. Two simulations similar to those shown in figure 26 except that $c = 3$; this implies that the prey species by itself would oscillate in numbers. In both cases the search area of adult predators is $\alpha = 0.005$ and of young predators is $\beta = 0.0005$. The difference between young and adult predators makes persistent coexistence possible (A) despite the fluctuations. In B there is cover for the prey ($x_c = 250$); note that this does not make for persistence.

For example, figure 27B shows simulation of a case identical to that in figure 27A, except that $X_c = 250$; since the carrying capacity $x_e = 5000$, this implies that 5% of the prey can be in cover. The effect is to convert a system which was oscillatory but persistent to one in which the predators die out.

There are other features which could stabilise a system of this kind. Territorial or hierarchical behaviour by the predators would be important. For example, Murton, Westwood and Isaacson (1964) have shown that populations of the Woodpigeon, *Columba palumbus*, are stabilised by hierarchical behaviour in the winter. Any process which eliminates some members of the population early in the winter in years when the population is high would help to prevent a crash. This emphasises the importance of differences between members of the population, of which age differences may be the most important.

5 COMPETITION

In this chapter I investigate the conditions which must be satisfied if two competing species are to coexist in the same habitat. In general, the treatment follows conventional lines. The intention is not to provide new insight into the problem, but to provide a background to the discussion in later chapters of more complex systems, with both competitive and predator–prey interactions.

A. Analysis based on the logistic equation

The usual analysis of competition between two species, going back to Volterra (1926), Lotka (1925) and Gause (1934), starts from the equations

$$\begin{aligned}
\dot{x} &= x(a-bx-cy), \\
\dot{y} &= y(e-fx-gy),
\end{aligned} \right\} \tag{36}$$

where a, b, c, e, f and g are positive constants. These equations are derived from the logistic equation, $\dot{x} = x(a-bx)$, by including the additional terms $-cy$ and $-fx$ to describe the inhibiting effects of each species on its competitor. It was suggested earlier that the logistic equation is best regarded as a purely descriptive equation, chosen as the simplest description of a population which increases exponentially when rare, and which approaches an equilibrium without oscillations. Therefore an analysis of equations (36) cannot tell us whether competitive interactions are likely to give rise to oscillatory behaviour, since they have been chosen to exclude that possibility. It is argued in section D of this chapter that competition will not cause oscillation in species which would not oscillate in its absence. Other limitations, arising from the particular form of equations (36) are discussed in section B.

The behaviour of the system described by equations (36) is best understood by plotting the lines $\dot{x} = 0$ and $\dot{y} = 0$ in the (x, y) plane, as was done when analysing continuously reproducing

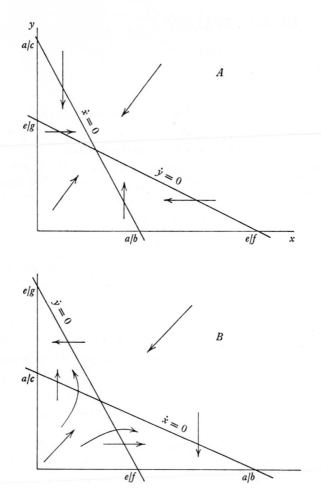

Figure 28. Competition between two species, showing A, stable and B, unstable equilibria.

predator–prey systems in Chapter 2. The two interesting cases, in which there is a non-trivial equilibrium, are shown in figure 28.

In case A, with $a/b < e/f$ and $e/g < a/c$, the equilibrium is stable. In case B, with both inequalities reversed, the equilibrium is unstable; which of the two species survives depends on the initial conditions.

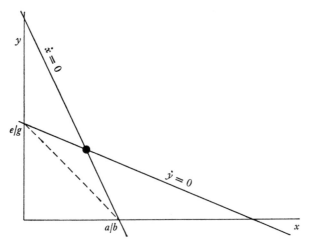

Figure 29. Competition between two species. The equation of the broken line is $x/x_e + y/y_e = 1$, where $x_e = a/b$ and $y_e = e/g$ are the equilibrium densities of the two species on their own. If the joint equilibrium is to be stable, corresponding to figure 28A, the equilibrium point must lie above this line.

There are two ways in which these inequalities can be interpreted in biological terms. First, suppose that the intrinsic rates of increase of the two species, a and e, are equal. Then the conditions for stability become $b > f$ and $g > c$. These inequalities state that an increase in numbers of either species inhibits its own growth more than it inhibits its competitor. If, for example, two species are limited, wholly or in part, by different resources, then the above inequalities are likely to hold; but if they have identical requirements, one of the two species is likely to be more efficient and will eliminate its competitor.

This is the basis of what has been called 'Gause's principle', or the principle of competitive exclusion. The principle has been variously stated; one formulation is that two species with identical requirements cannot coexist in a habitat. It has been pointed out by Hardin (1960) that Gause did not himself claim credit for the idea, ascribing it to Lotka (1925) and Volterra (1926). It was Lack (1947) who gave the credit to Gause; in the same book Lack himself gave a particularly clear formulation of the principle and showed how it could be used to explain the distribution of Darwin's finches on the Galapagos.

A slightly different interpretation of the stability criteria is illustrated in figure 29, corresponding to case A. The points $(a/b, 0)$ and $(0, e/g)$ are the equilibrium points for each species by itself. The condition for stability is that the joint equilibrium point should lie above the line joining these two points. The significance of this condition is more readily seen if we express the numbers of each species as 'equivalent numbers'; that is, as fractions of their equilibrium numbers, a/b and e/g respectively. If there is a stable equilibrium, then the total equivalent number at the joint equilibrium is greater than unity – that is, than the equivalent number at the single-species equilibria. This is understandable if the two species utilise different resources. It is, however, important to remember that this particular interpretation of the stability criteria depends, as will be shown in the next section, on the precise form of equations (36).

B. Competition with continuous reproduction – a more general case

The conclusion that case A in figure 28 is stable does not depend on the lines $\dot{x} = 0$ and $\dot{y} = 0$ being straight. In particular, they may be curved as in figure 30, so as to bring the equivalent numbers at the joint equilibrium below the equivalent numbers for the single species equilibria. That this can in fact happen has been shown by Ayala (1969) in experiments on population cages containing *Drosophila pseudoobscura* and *D. serrata*. Ayala originally interpreted his result as a disproof of the principle of competitive exclusion. As has been pointed out among others by Antonovics and Ford (1972) and by Gilpin and Justice (1972), it is more usefully thought of as illustrating the inadequacy of the logistic equation formulation to analyse all cases of competition.

Thus suppose that competition between two species is described by the equations

$$\left.\begin{aligned} \dot{x} &= xf(x,y), \\ \dot{y} &= y\phi(x,y). \end{aligned}\right\} \tag{37}$$

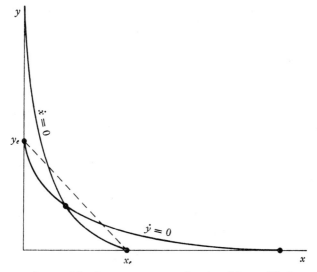

Figure 30. Competition between two species. A stable equilibrium which does not conform to the rule illustrated in figure 29.

We suppose also that a non-trival equilibrium exists with $f(x, y) = \phi(x, y) = 0$, and ask whether this equilibrium is stable.

The slope of the curve $f(x, y) = 0$ is given by

$$\frac{dy}{dx} = -\left[\frac{\partial f/\partial x}{\partial f/\partial y}\right],$$

and a similar expression gives the slope of the curve $\phi(x, y) = 0$. The condition for stability, from figure 28, is that the slope of the $\dot{x} = 0$ curve should be steeper than that of the $\dot{y} = 0$ curve where they intersect. That is,

$$-\left[\frac{\partial f/\partial x}{\partial f/\partial y}\right] < -\left[\frac{\partial \phi/\partial x}{\partial \phi/\partial y}\right],$$

or

$$\frac{\partial f}{\partial x} \cdot \frac{\partial \phi}{\partial y} > \frac{\partial f}{\partial y} \cdot \frac{\partial \phi}{\partial x}, \tag{38}$$

where the differentials are evaluated at the equilibrium point. Interpreting this inequality, the differentials on the LHS are the inhibiting effects of each species on itself, and on the RHS of each species on its competitor. They are measured for departures from equilibrium; an economist would call them the 'marginal' inhibiting effects. Note that (by the definition of competition) they are all negative, and hence that the slopes of the $\dot{x} = 0$ and $\dot{y} = 0$ lines are negative.

This conclusion was stated without proof by Gilpin and Justice (1972), in the form 'a necessary and sufficient condition for the stability of a competitive equilibrium is that the product of the intraspecific growth regulations be greater than the product of the interspecific growth regulations'.

It is worth asking what are the factors responsible for the curvature of the $\dot{x} = 0$ and $\dot{y} = 0$ lines implied by results such as Ayala's. The answer to this question is not known. One possible line of attack is as follows. Equations which would give a curvature as seen in figure 30 are:

$$\left.\begin{array}{l} \dot{x} = x(a - bx - cy - kxy), \\ \dot{y} = y(e - fx - gx - lxy). \end{array}\right\} \tag{39}$$

The terms in xy suggest that there is an inhibiting effect on the growth of both species which is zero when either species is absent, and maximum when both are common. A possibility which is worth investigating is that each species produces a substance toxic to the other, but only when the other is present. An alternative is that a substance toxic to both species is produced by one of the species, but only in the presence of the other. There is no evidence for the presence of such substances, but they would be worth looking for.

Notice that if each species produces a substance toxic to the other at a constant rate, independent of the density of the other, this is equivalent to introducing the terms $-ky$ into the first equation and $-lx$ into the second. This would convert case A (stable coexistence) into case B (victory for the species initially most abundant).

There are, of course, other functions which would give the curvature shown in figure 30, but equations (39) are the form which lend themselves most readily to biological interpretation.

C. Niche overlap and environmental variability

A weakness of the Lotka–Volterra model of competition is as follows: the model leads to the conclusion that coexistence of two species requires that there be a difference between the factors limiting them, but does not tell us how big a difference is necessary for persistent coexistence. Indeed the model suggests that a difference in requirements of an appropriate kind, however small, is sufficient to ensure coexistence. Common sense, however, suggests that, at least in a variable environment, persistent coexistence would require a difference of some definite magnitude. An important first step in deciding how big a difference is required to ensure coexistence in a varying environment has been made by May and MacArthur (1972). Their model will be described without going into mathematical details.

It is supposed that a series of species compete for a resource which varies along a single dimension; an example would be seed-eating species taking seeds of different sizes. Each species has a 'resource utilisation function' (figure 31), which in the case of seeds would be the proportion in the diet of seeds of different kinds. If the shapes of the resource utilisation functions are the same for all species, then the degree of competition can be measured by w/d, where w is the standard deviation of the resource utilisation function, and d the spacing between species.

If there are n species, with densities $N_1, \ldots, N_i, \ldots, N_n$, the competition equations, corresponding to (36), are

$$dN_i/dt = N_i\left[k_i - \sum_{j=1}^{n} \alpha_{ij}N_j\right]. \tag{40}$$

In these equations the constant k_i represents the suitability of the environment for the ith species – in the seed example, the abundance of seed of an appropriate size – and the coefficients α_{ij} measure the overlap in the utilisation functions of the ith and jth species. Thus the self-limitation term $\alpha_{ii} = 1$ for all species, and the α_{ij} will lie between 0 and 1, depending on the overlap between the species.

If the k's and α's are treated as constants, corresponding to a

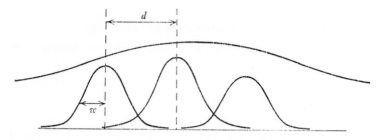

Figure 31. Upper curve: total resource available. Lower curves: resource utilisation functions for different species. The diagram (and the analysis of 'species packing' based on it) assumes that the species are competing for a resource which varies along a single dimension.

uniform environment, it is found that the species can be packed indefinitely close, short of complete congruence, and a stable equilibrium still exists. Environmental variability can be introduced into the model by putting

$$k_i = \bar{k}_i + \gamma_i(t),$$

where $\bar{k}_i$ is a constant mean value and $\gamma_i(t)$ is gaussian 'white noise', with variance σ^2. In the example of seed size, this implies random variations in the abundance of seeds of different sizes. It now turns out that community stability requires that d/w be greater than some specific value, which will depend on the noise level σ^2 and on the number of species and on other details of the model. The most interesting conclusion is that the value of d/w required for stability is not particularly sensitive to the noise level, over quite a wide range. Very approximately, for values of $\sigma^2/\bar{k}$ lying between 0.01 and 0.3, stability requires that d/w be greater than unity.

This conclusion does not depend critically on the shape of the resource spectrum or resource utilisation curves. It does, however, depend on the form of competition assumed. There are in effect two critical assumptions:

(i) competition is for a 'resource' which varies in a one-dimensional manner, and

(ii) if an item of the resource is taken by one individual it is not available to others.

The model seems to apply rather well to cases of competition for food items of varying sizes; it is less clear that it can be applied to cases in which one species replaces another with altitude, position on the shore, or other physical variables.

The importance of the model is that it does give a first clue to understanding species diversity.

D.   Absence of oscillations due to competition

It was shown in Chapter 2 that predator–prey interactions can cause oscillations, which may be divergent, in the numbers of a prey species which do not oscillate in the absence of predation. Competitive interactions do not produce oscillations. This is most easily seen from figure 28. Thus whatever the nature of the competitive interaction between two species, it is always the case (by definition) that an increase in either species reduces the growth rate of the other. That is to say,

$$\frac{\partial}{\partial y}\left(\frac{dx}{dt}\right) < 0; \quad \frac{\partial}{\partial x}\left(\frac{dy}{dt}\right) < 0. \tag{41}$$

It follows that both the $\dot{x} = 0$ and $\dot{y} = 0$ curves are monotonically decreasing, as shown in figure 28. Hence the equilibrium, if it exists, is either stable (case A) or unstable (case B), but in either case is non-oscillatory.

A similar conclusion follows if we consider the equations

$$\left.\begin{array}{l} X_{n+1} = X_n \phi(X_n Y_n), \\ Y_{n+1} = Y_n \psi(X_n Y_n), \end{array}\right\} \tag{42}$$

describing competition between species with discrete generations.

The equilibrium $\hat{X}\hat{Y}$ is given by the solution of the equations $\phi(\hat{X},\hat{Y}) = 1$; $\psi(\hat{X},\hat{Y}) = 1$. Writing $X_n = \hat{X}(1+x_n)$ and $Y_n = \hat{Y}(1+y_n)$, where x_n and y_n are small, it can be shown that

$$\left.\begin{array}{l} x_{n+1} = (1-b)\,x_n - cy_n, \\ y_{n+1} = -fx_n + (1-g)\,y_n, \end{array}\right\} \tag{43}$$

where
$$b = -\hat{X}\,\partial\phi/\partial X, \quad c = -\hat{Y}\,\partial\phi/\partial Y,$$
$$f = -\hat{X}\,\partial\psi/\partial X, \quad g = -\hat{Y}\,\partial\psi/\partial Y,$$

evaluating the differentials at the equilibrium. If the species compete, b, c, f and g are positive constants.

Eliminating x from equations (43) gives

$$y_{n+2} - (2-S)y_{n+1} + (1-S+R)y_n = 0, \qquad (44)$$

where $S = b+g$ and $R = bg - fc$. This equation has the solution $y_n = K_1\lambda_1{}^n + K_2\lambda_2{}^n$, where

$$\lambda_1, \lambda_2 = 1 - \tfrac{1}{2}S \pm \tfrac{1}{2}\sqrt{(S^2 - 4R)}. \qquad (45)$$

If $R < 0$, then $\lambda_1 > 1$, and y_n increases without change of sign. Hence a condition for stability is $bg > fc$; this corresponds to the 'Gilpin–Justice' criterion for the coexistence of continuously reproducing species.

Now $S^2 - 4R = (b-g)^2 + 4fc$, which is necessarily positive. Hence no long-period oscillations can occur. Oscillations with a two-generation period can occur if λ_1 or λ_2 is real and < -1. These short-period oscillations occur in the absence of competition (if either $b > 2$ or $g > 2$), and are not made more likely by competition.

6 MIGRATION

In this chapter I consider the effects of migration on the stability of a predator–prey system. I consider two types of situation. In the first, the habitat is more or less continuous, so that migration is a common event; in the second, the habitat is patchy, so that successful migration from patch to patch is relatively infrequent.

A. Migration in a continuous habitat

Consider a model with the following assumptions:

(i) The habitat is divided into a number of 'cells'. Within a cell the species can be regarded as single populations, spatial separation being ignored. Between neighbouring cells, migration is common, so that immigration and emigration have effects on the population size in a cell of the same order of magnitude as reproduction and death. Thus the model can represent either a continuous habitat, divided arbitrarily into cells for mathematical description, or a discontinuous habitat in which the discontinuities are not serious barriers to movement.

(ii) The effects of immigration and emigration are immediate. That is, the time taken to migrate is small compared to the life cycle.

(iii) Migration is 'conservative'; that is, there are no losses during migration, so that an individual leaving one cell arrives at another.

(iv) The environment is uniform, conditions in neighbouring cells being identical.

A model in which the first two of these assumptions are altered is considered in the next section. The effects of relaxing the last two assumptions are discussed briefly at the end of this section.

Suppose that the interactions between a predator and prey species are such that in a single cell, without migration, there is a large amplitude limit cycle. Our problem is then whether migration will reduce the amplitude of the limit cycle. It is clear that if the oscillations in two neighbouring cells are exactly in phase, then in view of assumptions (ii) and (iii) migration has no effect on the numbers in either cell, and therefore does not alter the amplitude of the cycle. Therefore the crucial question is whether migration tends to bring neighbouring cells into phase or to drive them out of phase. This question is answered by a method similar to that used by Winfree (1967) in analysing biological clocks.

The analysis is shown in figure 32. The behaviour of the system in a single cell is shown as a closed loop traversed in an anti-clockwise direction in the (x, y) plane, where x and y are the numbers of prey and predators respectively. Points 1, 2, 3 and 4 represent successive states of a cell A, and points 1', 2', 3' and 4' the corresponding states of a neighbouring cell B, assumed to be somewhat ahead of cell A in phase.

The arrows indicate the effects on the states of the two cells of migration between them, according to the following patterns of migration:

(i) Prey move away from cells with abundant prey.
(ii) Prey move away from cells with abundant predators.
(iii) Predators move away from cells with abundant predators.
(iv) Predators move away from cells with few prey.

These are net patterns of migration between neighbouring cells. In cases (i) and (iii), in which the movement of a species is a function of its own density, it makes no difference to the model whether the effect is due to 'diffusion' – that is, individuals are equally likely to move at any density so that there is a net movement away from cells with high density – or whether there is a density-dependent effect on the movement of individuals. Clearly, each of these patterns could be reversed, although in most cases reversal would seem unadaptive.

It is shown in figure 32 that patterns (ii) and (iv), in which the movement of one species is influenced by the abundance of the other, have no net effect on the phase difference between the two cells. Patterns (i) and (iii) tend to bring neighbouring cells into phase with one another.

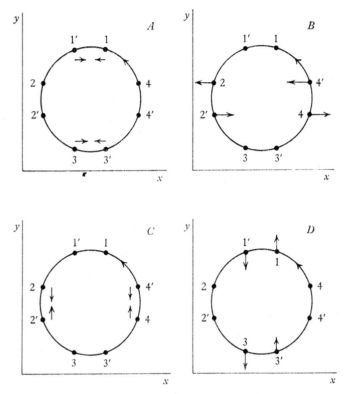

Figure 32. The effects of migration between neighbouring regions. x and y are the densities of prey and predator respectively. 1, 2, 3 and 4 represent successive states in one region, and 1′, 2′, 3′ and 4′ in a neighbouring region, assumed to be just ahead in phase. Arrows represent the effects of migration of the following kinds:

A prey move away from regions with abundant prey;
B prey move away from regions with abundant predators;
C predators move away from regions with abundant predators;
D predators move away from regions with few prey.

It is concluded that migration either by diffusion or by density-induced dispersal will tend to synchronise neighbouring cells. But it has already been shown that if neighbouring cells are synchronous, migration has no effect on the stability of the system. The only patterns of migration in which neighbouring cells will tend to become asynchronous are those in which either the predator or prey species migrates towards dense associations of its own species.

If migration is not conservative, so that substantial numbers of individuals are lost on migration, this will not alter the conclusion that neighbouring cells will become synchronised. The effect of such migration will be equivalent to a reduction in the reproductive rate within a cell of the species concerned. This will tend to reduce the amplitude of oscillations. However, migration of this type is difficult to explain in evolutionary terms, unless the migrants would in any case be excluded from reproduction if they stayed put.

If the environment in neighbouring cells is different, then the natural period of the limit cycles will be different. Provided the differences are not too great, neighbouring cells will be brought into phase by migration, with a period intermediate between their natural periods. Two-cell systems of this kind have been simulated. The effect is to reduce the amplitude of oscillation, but the effect is small, and unlikely to be important in comparison to the stabilising effects of limitation by a fixed resource and of cover for the prey species.

B.   Migration and local extinction

The conclusion of the previous section that migration cannot stabilise an intrinsically unstable predator–prey system seems to be contradicted by Huffaker's experiments (1958). These experiments concerned a predatory mite *Typhlodromus occidentalis* and its prey, a phytophagous mite *Eotetranychus sexmaculatus*. The environment consisted of a number of oranges which formed the food for the prey species and which were replaced from time to time. The number of oranges, and the ease with which the mites could migrate from one orange to another, were varied. In most cases there was a rise and subsequent fall in the abundance of the prey, closely followed by a corresponding rise and fall in the abundance of the predator and by the extinction of the predator. In order to get continued coexistence of the two species through three cycles of abundance (figure 33), it was necessary to increase the number of oranges to 120, to slow down the migration of the

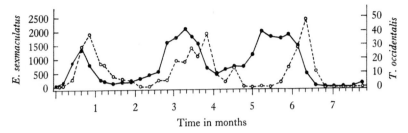

Figure 33. Three successive cycles of abundance of a laboratory population of a herbivorous mite (full line) and a predatory mite (broken line), from Huffaker (1958).

predators by inserting vaseline barriers, and at the same time to increase the migration of the prey by providing them with wooden posts from which they could launch themselves suspended on a silken strand, aided by currents from an electric fan.

The models considered in this section were suggested by Huffaker's experiments. Their essential feature is as follows: Any local patch of environment (*e.g.* an orange) – referred to below as a 'cell' – which contains both prey and predators will at some time in the future become 'empty'; that is, both predator and prey will become extinct. The cell is then available for fresh colonisation.

In the simplest model of such a system, a cell can be in one of three states:

E   empty, containing neither prey nor predators;
H   containing prey ('herbivores') only;
M   containing both prey and predators (mixed).

I will at first make the simplifying (and, as it will turn out, misleading) assumption that state transitions occur simultaneously in all the cells of a system. Thus we take as unit time T the period from the invasion of an H cell by a predator to the time when that cell becomes empty. At intervals of T, all cells in the system may undergo state transitions, with the following probabilities:

	E	H	M
E	$1-PH$	PH	0
H	0	$1-PP$	PP
M	1	0	0

In this table, some justification is needed for the assumption that the transition $E \rightarrow M$ is impossible. It is supposed that a predator arriving in a cell by migration will starve unless there is already an adequate population of herbivores. Thus an E cell can be converted into an M cell only after an intervening period as an H cell, during which the herbivore population is building up.

In this table, PP is the probability that a cell will be invaded by at least one predator in the time T, and PH that it will be invaded by at least one prey. In general, PH and PP will be variables, since they will depend on how many neighbouring cells contain prey and predators respectively.

How will such a system behave? I will consider only two simple cases, since it will become apparent that the 'synchrony' assumption is sufficiently misleading to make a more detailed analysis not worthwhile.

Consider first a model with many cells, and a pattern of migration such that any cell can be reached from any other; such models have been called 'island' models in population genetics. I will also suppose that the number of migrants is such that $PH = PP = 1$. Each individual cell then goes through a series of transitions $E \rightarrow H \rightarrow M \rightarrow E \ldots$ indefinitely. The time averages of the total number of E, H and M cells are therefore equal. There will, however, be a three-generation cycle in the numbers of each type, whose amplitude will depend on the initial conditions. Thus taking E, H, and M to refer to the numbers of cells at any instant in the corresponding states, and supposing that in a system consisting of 50 cells the initial values are $E = 25$, $H = 20$, $M = 5$, the subsequent behaviour would be:

Time interval	0	1	2	3	4	5	...
E	25	5	20	25	5	20	
H	20	25	5	20	25	5	
M	5	20	25	5	20	25	

If the probability of migration is low, so that either PH or PP is substantially less than unity, then the time averages of the numbers of cells in the three states are no longer equal. In a finite system of this kind, one or both species will ultimately go

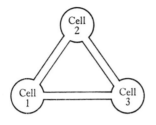

$$\begin{matrix} & E & & H & & M & & E \\ M & & H \to E & & M \to H & & E \to M & & H \end{matrix} \cdots$$

Figure 34. An environment with three cells.

extinct, although the expected life of the system, which depends on *PH*, *PP* and on the number of cells, may be very great.

At first sight, this model might suggest that provided both prey and predator have sufficient power of migration, a system of this kind will last indefinitely. However, it is easy to see that this conclusion cannot be correct, since if powers of migration are sufficiently increased, all the cells of a system can be regarded as a single large cell, in which rapid extinction is certain. The weakness of the model lies in the assumption of synchrony, as can be seen by considering a system consisting of three cells only, as in figure 34.

According to my assumptions, if $PH = PP = 1$, this 3-cell system will go through the three-generation cycle shown in the diagram indefinitely. But if I abandon the synchrony assumption, this persistence breaks down. Thus suppose that at time $t = 0$ the system state is *MEH*, the *M* cell having just been invaded by a predator. On the synchrony assumption, the *E* cell must now remain empty for a time *T*. But in practice, at time δt a herbivore might reach the *E* cell from the *H* cell, giving a system state *MHH*. Subsequent transformations would then be $MHH \to EMM \to EEE$.

A more realistic model

A completely realistic model would have to abandon the description of a cell as being in one of a small number of discrete states, and instead characterise it by the numbers, ages, etc., of the prey and predators it contained. Although realistic, such a model would be difficult to analyse or understand. Instead, I retain a discrete-state model with synchronous state transitions, but increase the number of possible states of a cell, so as to reduce the distorting effects of the synchrony assumption. This is analogous to replacing a differential equation by a finite difference equation in which the time intervals have been made sufficiently short to ensure that its behaviour resembles that of the original equation. The hope is that the model reflects reality well enough to give some insight into the effects on persistence of parameters such as relative migrating ability of prey and predator, the existence of cover for the prey, and the number and connectivity of the cells.

The model is shown in table 1 and figure 35. The states are as follows:

E	empty;
HA	few prey;
HB	increasing prey;
HC	many prey;
MA	many prey, few predators;
MB	many prey, increasing predators;
MC	many prey, many predators;
MD	few prey, many predators.

TABLE 1. *States and transition probabilities in a model of a predator–prey system with immigration and local extinction*

	E	HA	HB	HC	MA	MB	MC	MD
E	$(1-PH)$	PH	0	0	0	0	0	0
HA	0	0	$1-PP$	0	PP	0	0	0
HB	0	0	0	$1-PP$	PP	0	0	0
HC	0	0	0	$1-PP$	PP	0	0	0
MA	0	0	0	0	0	1	0	0
MB	0	0	0	0	0	0	1	0
MC	0	0	0	0	0	0	0	1
MD	1	0	0	0	0	0	0	0

It is assumed that the numbers of migrating individuals is always small. Hence migrants are responsible for the transitions $E \rightarrow HA$ and $H \rightarrow MA$, but do not alter significantly the rates of the transitions $HA \rightarrow HB \rightarrow HC$ or $MA \rightarrow MB \rightarrow MC \rightarrow MD$.

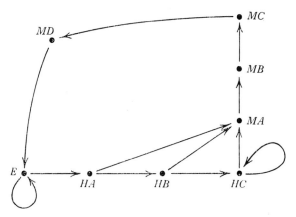

Figure 35. Graph showing transitions in a system with eight states. The graph has been oriented so as to represent the predator–prey cycle in state space, comparable to the cycles in figure 32.

A migrant predator arriving in an E cell starves, but can convert an HA, HB, or HC cell into an MA cell. A migrant herbivore converts an E cell into an HA cell.

The probability PH that an E cell will be converted into an HA cell is $1-(1-ph)^r$, where ph is the probability that, in unit time, a herbivore will reach the E cell from a neighbouring cell which contains many herbivores, and r is the number of neighbouring cells containing many herbivores – that is, which are HC, MA, MB, or MC. In any particular simulation, ph is constant and r a variable.

Similarly PP, the probability that an HA, HB, or HC cell is converted into an MA cell is $1-(1-pp)^s$. Here pp is the constant probability that a predator will reach the cell from a neighbour containing migrant predators. Two versions of the model have been considered. In one version it is supposed that predators only migrate when they are hungry, so that s is the number of neighbouring MD cells. In the others, s is the number of MB, MC, and MD cells.

The behaviour of the model has been analysed by computer simulation for different values of ph and pp, for 'island' and 'stepping stone' models and for varying numbers of cells. The general conclusion is that such a model can rather easily give

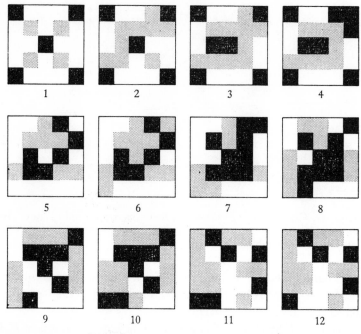

Figure 36. Simulation of the system shown in figure 35, in an environment of 25 cells; showing the first twelve time intervals. Black squares represent cells with both predator and prey, hatched squares cells with prey only, and white squares empty cells.

persistent coexistence of predator and prey; that is, persistence does not call for a particularly careful choice of parameters.

I will describe first a case for which the two species persist. The 25 cells are arranged in a square 5 × 5 array. The model is a 'stepping stone' model, migration being possible only between a cell and its four closest neighbours. The 'second version' is considered, in which predators migrate from MB, MC, and MD cells. Migration probabilities $ph = pp = 0.438$.

This model was run five times, with different initial conditions, for 100 time intervals on each occasion, without either species going extinct. The numbers of E, H, and M cells remain fairly steady, fluctuating either side of the theoretical equilibrium values calculated below. Figure 36 shows a short section of one such run.

A number of computer simulations lead to the conclusion that permanent coexistence is favoured by the following features:

(i) Prey with a high capacity for migration (*PH* large).

(ii) Cover or refuge for the prey. If it is supposed that a few prey survive the extinction of the predator, this is equivalent to assuming that E cells are always converted into HA cells – that is, $PH = 1$.

(iii) Restriction of the period during which predators migrate. Thus version 1, in which migration occurs only from MD cells, is more permanent than version 2.

(iv) A large number of cells. Clearly, the smaller the number of cells, the greater the chance of extinction.

It will be seen that the illustrative example does not have all these properties.

The ability of predators to migrate must not be too low, or they will go extinct; a high capacity to migrate favours permanent coexistence so long as the prey are also highly mobile. However, it must be remembered that it is an assumption of the model that immigration into a cell is always negligible compared to reproduction in that cell (provided the latter is not zero).

It is difficult to generalise about the effects of connectivity of cells – that is, to compare the permanence of island and stepping stone models. Thus

$$\widehat{PH} = 1 - (1 - ph)^{\hat{r}},$$
$$\widehat{PP} = 1 - (1 - pp)^{\hat{s}},$$

where $\widehat{PH}$, $\widehat{PP}$, $\hat{r}$ and $\hat{s}$ refer to the values at equilibrium. Thus for a stepping stone model $\hat{r}$ and $\hat{s}$ lie between 0 and 4, whereas for an island model they may be large, depending on the total number of cells. It therefore seems best to compare the persistence of island and stepping stone models while keeping $\widehat{PH}$ and $\widehat{PP}$ constant, and varying ph and pp accordingly. That is, we compare the persistence of the two types of model when the total probability of a migrant arriving, from any other cell, is kept constant (at equilibrium). This implies much lower values of ph and pp for the island models.

With this assumption, the island model is more persistent than the stepping stone model, but the difference is not very great.

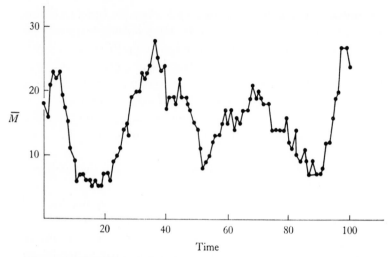

Figure 37. A long-period cycle in a simulation of the system shown in figure 35. This is an 'island' model with 100 cells, and $\widehat{PH} = \widehat{PP} = 0.1$ (for definitions see p. 74). $\overline{M}$ is the total number of cells containing predators.

This suggests that in nature persistence would be favoured if occasional individuals can migrate long distances.

One last conclusion from the computer simulations will be mentioned before giving some analytical support. This concerns the oscillatory behaviour of the model. Short-period oscillations, of two, four or eight generations, are observed in the 'island' model. These resemble the three-generation cycles described above for the simple three-state model; they probably represent a weakness of the model rather than corresponding to anything that happens in real ecosystems. However, long period oscillations may also occur, as shown in figure 37. These appear not to be a product of the assumption of a series of discrete states with synchronised transitions. As is usual in ecological oscillations, they are most easily understood as being caused by delays in regulation, arising because of the time elapsing between the moment when a cell is invaded by herbivores or predators and the time when it becomes a source of invaders. The fluctuations shown in figure 37 should be compared to those found by Huffaker (figure 33).

Some insight into these conclusions can be obtained as follows: let E, HA, HB, HC, MA, MB, MC, MD be the fraction of cells in the corresponding states at equilibrium, and let $\widehat{PH}$ and $\widehat{PP}$ be the values of PH and PP at equilibrium. Further, let $HA + HB + HC = H$, and $MA + MB + MC + MD = M$. We want to find values of E, H and M, the proportions of empty, herbivore and mixed cells at equilibrium, as functions of $\widehat{PH}$ and $\widehat{PP}$. Referring to table 1,

$$MA = MB = MC = MD,$$
$$E(1 - \widehat{PH}) + MD = E, \quad \text{or} \quad E = MD/\widehat{PH},$$
$$HA = E.\widehat{PH} = MD,$$
$$HB = HA(1 - \widehat{PP}) = MD(1 - \widehat{PP}),$$
$$HC(1 - \widehat{PP}) + HB(1 - \widehat{PP}) = HC, \quad \text{or} \quad HC = MD(1 - \widehat{PP})^2/\widehat{PP}.$$

Hence
$$E = MD/\widehat{PH},$$
$$M = 4MD,$$
$$H = MD/\widehat{PP},$$

and since $E + M + H = 1$, $MD = 1/(4 + 1/\widehat{PH} + 1/\widehat{PP})$.

Hence for any values of $\widehat{PH}$ and $\widehat{PP}$, it is possible to find the equilibrium fractions E, H and M. Some values are given in table 2.

TABLE 2. *Equilibrium frequencies of empty cells, cells with herbivores, and cells with herbivores and predators, as a function of $\widehat{PH}$ and $\widehat{PP}$, the probabilities that a cell will receive at least one immigrant herbivore, or predator, in unit time*

	$\widehat{PP} = 0.1$			$\widehat{PP} = 0.9$		
	$\hat{E}$	$\hat{H}$	$\hat{M}$	$\hat{E}$	$\hat{H}$	$\hat{M}$
$\widehat{PH} = 0.1$	0.417	0.417	0.166	0.662	0.073	0.265
$\widehat{PH} = 0.9$	0.073	0.662	0.265	0.178	0.178	0.644

For such a system, extinction may occur for one of three reasons:

(i) All the H cells may be invaded by predators and may go to extinction before any E cells have been invaded by herbivores. Both species extinct.

(ii) All the M cells may become E cells before any predators have migrated from them to H cells. Predators only extinct.

(iii) The equilibrium may be unstable, and there may be large amplitude oscillations. This will make both the first two types of extinction more likely.

An examination of the simulations, together with some very approximate calculations based on the equilibrium values in table 2, lead to the following conclusions.

Extinction of type (i) is most likely when $\widehat{PP}$ is large and $\widehat{PH}$ small. In this case, since H is small and $\widehat{PP}$ large, no cells with herbivores only escape invasion for long. Consequently, even though E is large, the low value of $\widehat{PH}$ means that there is a danger of extinction.

Extinction of type (ii) is most likely when both $\widehat{PP}$ and $\widehat{PH}$ are small. In this case, the low value of $\widehat{PP}$ leads to a high probability that no H cell will be invaded. Extinction is more likely for low than for high $\widehat{PH}$ because there are fewer H cells at risk.

An important question, particularly relevant to problems of biological control, concerns the density of the herbivores at equilibrium, as compared to their density in the absence of predation, which would be 1.0. The table shows that an equilibrium can exist with the herbivores density much less than the carrying capacity. This requires PH small and PP large. Simulations showed that with small numbers of cells (20–25), and with these values of PH and PP, extinction of one or both species occurred rapidly. A deterministic 'island' model, assuming large numbers of cells, did give a persistent equilibrium with a low herbivore density and most cells empty.

Finally, it has to be explained why the argument of the last section does not apply to the models considered here. Figure 35 shows the states of a cell in a phase diagram. The important point to note is that a cell on its own will not traverse this cycle; instead, it will come to rest at state E or state HC, depending on its initial state. Thus a cell will tend to keep out of phase with its neighbour, pausing at E or HC until its neighbour is out of phase. Thus a cell will pause at E until its neighbour reaches HA, MA, MB or MC, and at HC until its neighbour reaches MD (in version 1) or MB, MC or MD (in version 2). Therefore the argument that neighbouring cells will come into phase does not apply in this case.

C. The evolution of migration

If the habit of migration, as opposed to staying put, has evolved, it will usually be because there has on the average been a selective advantage to those individuals who migrated. The alternative is that migration has evolved by kin selection; that is, although migration reduced the expected number of progeny of migrants, it increased the expected number of progeny of close relatives of those migrants. Although kin selection cannot be ruled out, it is ignored in the discussion which follows.

The selective advantage of migration will depend on the nature of the habitat, as follows:

1. Habitat discontinuous and transitory

Examples are recently disturbed earth (for colonising plants), host organisms (for their parasites), dung and corpses (for decomposing organisms). In such habitats some dispersal mechanism is a necessity for survival.

2. Habitat discontinuous and long-lasting

In any patchy environment, the duration of an environment containing at least some suitable patches is likely to be much greater than the duration of any individual patch. Consequently even if patches are long-lasting compared to the life span of an individual organism, some readiness to migrate will be favoured by selection. The extent of migration will depend on the population dynamics within a patch. At one extreme, if population numbers within a patch are uniform, migration will rarely confer selective advantage. At the other, if populations frequently go extinct, as in the models considered in the preceding section, migration will frequently confer selective advantage.

3. *Habitat continuous*

In this case, most movement will be short-range and diffusive. If population numbers oscillate with large amplitude, neighbouring populations will oscillate in phase, although distant populations may be out of phase. Therefore at a time of over population, short-range migration will confer no selective advantage, since the migrant individual will move to another overpopulated region. Selection may, however, favour a pattern of migration in which individuals from populations at peak density migrate long distances, and cross barriers not crossed in normal diffusion movements. Lemming migrations are a possible example of such a pattern.

7 STABILITY AND COMPLEXITY – AN INTRODUCTION

In their primer on population biology, Wilson and Bossert (1971) remark that it is part of the 'conventional wisdom' of ecology that the observed stability of natural ecosystems is to be explained by their complexity – that is, by the great number of species composing them and of interactions between them. The reasons for thinking that increasing complexity leads to stability were clearly summarised by Elton (1958). Simple laboratory models of predator–prey systems, containing only two or a few species, are usually unstable. Outbreaks are commoner in cultivated land than in natural communities, and are more serious when only one crop is grown. At the other extreme, outbreaks do not occur in tropical forests, and numerical stability is much greater in species-rich forests than in the sub-arctic with its relatively few abundant species. Similar views were expressed by Hutchinson (1959).

These observations do support the view that complexity leads to stability, although other explanations are possible. For example, differences between natural and agricultural communities may be caused by the fact that the species in a natural system have undergone long periods of coevolution, and the difference between tropical and sub-arctic systems by the destabilising effects of climatic fluctuations on the latter. The remainder of this book is concerned with a theoretical analysis of the relation between stability and complexity. The subject is difficult, and the conclusions reached are tentative. The main structure of the argument to be followed is briefly set out in the remainder of this chapter.

A. The method of statistical mechanics

This approach has been pioneered by Kerner (1957, 1959) and Leigh (1965, 1968), and is summarised in Chapter 8. Despite the elegance and analytical power of the method, I have doubts about its utility in ecology. The fundamental difficulty is as follows. In order to apply the method at all, it is necessary to have a set of equations of motion for which it is possible to find a 'conserved quantity' – comparable to the total kinetic energy of the molecules in a perfect gas. For most of the equations of ecology (for example, for the logistic equation) no such conserved quantity exists. Consequently Kerner and Leigh have started from Volterra's predator–prey equations without damping. The artificiality of such equations has already been discussed (p. 11). They lead to a structurally unstable dynamic system; that is, any small change in the form of the equations (by adding $\epsilon\phi(x)$, where $\phi(x)$ is an arbitrary function and ϵ is small) would completely alter the behaviour of the system.

B. Complexity at a single trophic level

This topic is discussed in Chapter 9.

It is a well-known result (see p. 60) that if two competing species are to coexist, two inequalities must be satisfied. As the number n of competing species increases, the number of inequalities which must be satisfied by the parameters describing the competition will likewise increase. It has in fact been shown by Strobeck (1973) that the number of inequalities is $2(n-1)$. It is clear that the chance of coexistence when n is large is very small, unless the parameters themselves have values which are highly non-random.

I then consider the problem of how many competing species can coexist, as a function of the number of resources available to them. The basic conclusion, due to MacArthur and Levins (1964) and to Rescigno and Richardson (1965), is that there cannot be more species in a habitat than there are resources on which they depend. There are two difficulties with this theorem. The first is that it states only that, with n resources, there cannot be more than n dependent species; whether there will be as many as n, and indeed whether there will be more than one, depends on evolutionary criteria, to be discussed in Chapter 11. The second difficulty lies in the definition of a 'resource'. I attempt to clarify this a little further.

I also consider an extension of the theorem by Levin (1970). This introduces two new features. First, it introduces climate and other non-density-dependent factors as limiting factors. I give reasons for thinking that this is wrong, and that it leads to irrelevant complications. Second, and more important, it incorporates predator–prey as well as competitive interactions, and leads to a rather general formulation of the conditions (necessary, but unfortunately not sufficient) which must be satisfied if two species are to coexist in a habitat.

C. Complexity in ecosystems with several trophic levels

In Volterra's description of a predator and its prey, with the prey resource-limited (equations (8)), coexistence requires that a single inequality be satisfied. It is easy to show that if the food chain is lengthened to n steps, a single inequality is still sufficient to ensure the coexistence of all $n + 1$ species. Thus there is no argument, comparable to the argument above concerning n competing species, why an increase in the number of trophic levels in a system should in itself lead to instability.

I consider two lines of attack on the stability of many-species ecosystems. The first, pioneered among others by May (1971 a, b, 1972), is to consider the stability of ecosystems described by systems of differential equations, in which the signs and magnitudes of the interaction coefficients are varied randomly. For a rather general type of equation, May is able to show that an increase in complexity, either in the number of species or in the number of interactions between them, tends to cause instability of the system as a whole. This leads very forcibly to the conclusion that if complex natural ecosystems are stable, it can only be because the interactions occurring in them are highly non-random.

An alternative line of attack is to consider a model ecosystem in which the magnitude of the interactions is altered systematically rather than randomly. This is done for an ecosystem described by finite difference equations instead of differential equations, in order to introduce the important destabilising effects of time delays. This analysis leads to the conclusion that competition for resources at the lower trophic levels of an ecosystem leads to instability, but that competition between the top predators has a stabilising effect.

D. The evolution of ecological parameters

It is clear from the preceding discussion that the stability of complex ecosystems requires that the parameters describing the interactions between species must be highly non-random. These parameters, or rather the interactions they describe, are the product of natural selection. It is this fact which is the main justification for a science of 'population biology' which embraces both ecology and population genetics.

It would be a serious fallacy to suppose that selection can act effectively by favouring the survival of stable ecosystems rather than unstable ones. Selection preserves and promotes characteristics which ensure the survival and reproduction of the individuals which possess them; it can occasionally, by kin selection, promote characteristics which ensure the survival of close relatives of that individual, and even, by group selection, those which ensure the survival of large conspecific populations. But it is not reasonable to think of an ecosystem as a whole as a unit upon which selection can act.

There are two main ways in which selection might act to produce complex and stable ecosystems, which I will call 'genetic feedback' and 'species exclusion'.

If two species do coexist for a time in an ecosystem, and if they interact, then they are likely to undergo genetic changes caused by this interaction; that is, they will 'coevolve'. It follows that the parameters in our ecological equations will be modified by natural selection. They need not evolve in such a way as to stabilise those equations; whether they do so will depend in the main on the selective advantage to the individual. There are some situations in which there are good reasons to think that selection will usually have stabilising consequences. An example is the evolution of parasites from a lethal to a more nearly commensal relation to their hosts. It is worth noting that there may be an element of kin or group selection in this process; by refraining from damaging the host, a parasite is ensuring a more lasting home not only for itself but for the other parasites in the same host, which will often be close relatives. Pimentel (1968) has argued that there are more general reasons why coevolution should usually lead to ecological stability, a process he has called 'genetic feedback'. The validity of his argument is considered in Chapter 11.

The process of 'species exclusion' can most easily be imagined by considering the fauna or flora of an island. Immigrant species will from time to time reach the island. Such species may fail to establish themselves. A species which does establish itself may be a simple addition to the community, or it may eliminate one or more of the pre-existing species through predation or competition; it may even eliminate other species and then itself become extinct. The species on the island at any time will form a reasonably persistent ecosystem. They are, however, in no way a random sample of the species which have at one time or another reached the island; they are in fact a sample selected because of their ability to coexist. According to this argument, actual ecosystems, complex or otherwise, are persistent because species whose presence would be inconsistent with persistence have been excluded. The argument can easily be applied to island communities, but its relevance to the communities on continental land masses is less clear.

In Chapter 11 I consider the selective forces which operate when two species coexist in the same habitat. I first consider whether there are any general reasons why coevolution should usually lead to ecological stability. I then consider the selective forces tending to make a species into a specialist or a generalist. This has an obvious relevance to the stability of ecosystems, because two species specialising on different resources will coexist whereas two competing generalists may not. In Chapter 12 I consider the evolution of territorial behaviour, since this can lead to immediate self-regulation of predators, and therefore have important stabilising effects.

8 THE STATISTICAL MECHANICS OF POPULATIONS

One possible approach to the analysis of complex ecosystems is to use the techniques of statistical mechanics, developed by physicists to analyse systems too complex to tackle by the more orthodox approach of finding solutions to deterministic equations. The aims and prospects of this approach to ecology will be easier to understand in the light of what has actually been achieved by the same approach in physics. In essence, statistical mechanics attempts to explain known laws connecting 'macroscopic' variables such as temperature or pressure (*e.g.* the 'gas law', $PV = MRT$) in terms of the motion of 'microscopic' particles obeying the known laws of motion of mechanics (*e.g.* force = mass × acceleration).

Since there are of the order of 10^{23} particles in a handlable quantity of gas, it is clearly out of the question to solve the equations of motion explicitly. But it has proved possible to make statements about statistical properties of populations of particles: examples of such statements are the 'equipartition' theorem, which states that in a gas containing particles of different mass, the kinetic energy, averaged over time, is the same for all particles, or the Maxwell–Boltzmann law, which gives the velocity distribution for a number of particles of the same mass. Ecological analogues of these two statements are described below. Statistical mechanics also makes statements about macroscopic observables. For example, the pressure is equal to $2nmu$, where n is the average number of particles, of mass m, colliding with unit area of wall per unit time, and u is the average velocity normal to the wall prior to (or after) the collision. These achievements depend on finding some conserved quantity which remains constant throughout the motion; the ecological application also depends on finding such a quantity.

In ecology the microscopic variables, corresponding to the mass or velocity of a particle, have been taken to be the population densities of individual species. A 'closed' ecological system – corresponding to a closed physical system to which there is no input or output of matter or energy – clearly cannot be closed in the physical sense, but is taken to be one with no immigration or emigration. The fact that in the real world there are no closed ecosystems (other, perhaps, then the whole planet) need not worry us too much, since there are no closed physical systems either.

It is less clear what should be the 'macroscopic' variables of ecology. Possible macroscopic variables are the primary productivity, the biomass, the number of species, the mean and variance of the number of individuals per species, and the mean amplitude of fluctuations in the numbers of individuals. An example of a law connecting macroscopic variables, corresponding to the gas law, is the 'law' $N = AS^k$ connecting the number of species N in a given taxon on an island with the area S of the island. However, there are not in ecology a set of macroscopic variables generally accepted as appropriate for the description of ecosystems, as pressure and temperature were accepted as appropriate to the description of gases before they were given any microscopic interpretation. It is therefore one hope of those responsible for applying these methods in ecology that such macroscopic variables might be discovered or invented.

A second difficulty lies in the absence of anything corresponding to Newton's laws of motion. For reasons to be discussed in a moment, the choice of formulation of the microscopic laws of motion has fallen on Volterra's equations without damping. These equations cannot claim to have as close a correspondence to reality as Newton's laws, or as Mendel's laws which form the basis for the mathematical development of population genetics.

There is one respect in which the development of a statistical mechanics of ecology is easier than the corresponding theory of gases. We can sometimes follow directly the change in a microscopic variable – *i.e.* the density of a single species. It is as if a physicist lacked a concept of temperature, and had no form of thermometer, but nevertheless hoped to reach a meaningful concept of temperature by following changes in the velocity of individual particles in a gas. It is perhaps worth noting that if it were possible to follow the motions of individual particles in this way, one might well reach a concept of temperature, and thence be led to invent an instrument to measure it.

The development of a statistical mechanics of ecology is due in the first instance to Kerner (1957), and subsequent elaborations have been made by Kerner (1959) and Leigh (1965, 1968). In the account that follows, I describe the assumptions they make, and give some of the more interesting conclusions they reach, without attempting to give the mathematical derivation of those conclusions. I then discuss the validity of their assumptions, and consider various ways in which their conclusions might be tested observationally.

Consider a closed ecosystem containing k species, with numbers N_1, N_2, ..., N_k. It is assumed that the change in time of these numbers is fully described by the equations

$$\frac{\mathrm{d}N_r}{\mathrm{d}t} = \epsilon_r N_r + \frac{1}{\beta_r} \sum_s \alpha_{sr} N_s N_r. \tag{46}$$

These are Volterra's equations, ignoring time delays due to breeding seasons or age structure. It is assumed that 'self-limiting' or 'damping' terms of the form $\alpha_{rr} N_r^2$ are absent – *i.e.* that $\alpha_{rr} = 0$ for all r. It is further assumed that $\alpha_{sr} = -\alpha_{rs}$ for all species pairs. Thus all species interactions are of the predator–prey type; there are no competitive or commensal interactions. The reason for omitting all self-limiting and competitive effects is that only by so doing is it possible to obtain a system with a conserved quantity, and as was explained above, some conservation law is needed if the methods of statistical mechanics are to be applied.

The assumption $\alpha_{sr} = -\alpha_{rs}$ implies that each prey individual eaten is converted into one predator, which is not generally the case. This difficulty can be removed as follows. First, the N_r are taken as the biomass of a species rather than its numbers. Second, if λ units of prey biomass are required to produce one unit of predator biomass, then in equation (46) the N_r for predator species should be λ times the actual biomass. In an ecosystem with several levels, the species at the third level should be weighted by $\lambda_1 \lambda_2$, where λ_1 is the conversion efficiency between the first and second level, and λ_2 between the second and third level, and so on.

If the equilibrium value of N_r is q_r, then

$$\epsilon_r \beta_r + \sum_s \alpha_{sr} q_s = 0. \tag{47}$$

In what follows it is assumed that equations (47) have a unique solution, with all q's positive; necessary (but not sufficient) conditions for this are that k be even, and that all ϵ's be not the same sign.

Volterra has shown that the N_r are variable between finite positive limits; some, and often all, fluctuate without damping, and their time averages are the equilibrium values q_r.

To find a conserved quantity, it is convenient to change the notation, putting $N_r/q_r = n_r$, and $v_r = \log n_r$. The basic equations then become

$$\frac{dv_r}{\beta_r dt} = \sum_s \alpha_{sr} q_s (\exp v_r - 1). \tag{48}$$

It can then be shown that

$$G = \sum_r \beta_r q_r (\exp v_r - v_r) = \text{const.} \tag{49}$$

Thus G is a constant of the motion. Note that it is a sum of a number of terms $G_r = \beta_r q_r (\exp v_r - v_r)$, one corresponding to each species. Each G_r has a minimum $\beta_r q_r$, when $v_r = 0$ and hence $N_r = q_r$, its equilibrium value. To say that G is a constant of the motion implies that if the system is started off with $G = G_0$, then the value G_0 will be maintained indefinitely. Of course, the same system (*i.e.* a system with the same values of ϵ_r, β_r, α_{sr}) can start off with a high or low value of G. In general, the further the initial values of N_r are from their equilibrium value q_r, the greater G.

Using the fact that G is conserved, Kerner defines a modulus θ, which he calls the 'temperature' of the system, and shows that for all r species

$$\frac{\beta_r}{q_r} \overline{(N_r - q_r)^2} = \theta, \tag{50}$$

where the bar indicates that a time average is to be taken of $(N_r - q_r)^2$. Thus the temperature θ measures the mean square deviation of each species from its equilibrium number. The fact that (50) applies to all species implies that all species fluctuate with equal 'energies'; thus (50) is an analogue of the statement in mechanics that all particles have the same average kinetic energy.

Kerner then finds the probability that the numbers of a species at any instant lie in the interval n_r to $n_r + dn_r$, namely

$$P(n_r)\, dn_r = \text{const.}\ n_r^{x_r-1} \exp\left(-x_r n_r\right) dn_r, \qquad (51)$$

where $x_r = \beta_r q_r/\theta$. This is the analogue of the Maxwell–Boltzmann distribution. Kerner points out that if θ is very large, (51) reduces to the distribution which Fisher (in Fisher, Corbett and Williams, 1943) was led to assume for the numbers n_r of different species in an ecosystem, in order to account for the observed numbers of individuals of different species of tropical moths in a sample. The relevance of this agreement between theory and observation is discussed below.

Further deductions from Kerner's model have been made by Leigh (1965, 1968). In particular, he considers the 'stability' of the system; by stability he means approximately what I have called 'persistence' in this book, since he is concerned with the frequency with which a species undergoes gross fluctuations which, in a finite system, might lead to extinction. He uses the variable $x_i = (N_i - q_i)/q_i$; thus x_i is a normalised departure from equilibrium. The mean value of x_i is zero. Leigh asks how frequently the value of x_i departs from its zero equilibrium value by exactly a given distance a. Clearly, for a given θ, the greater a the less frequent will be such a departure. Leigh shows that the frequency is

$$\frac{1}{\pi} \exp\left(-\frac{q_i a^2}{2\theta}\right)\sqrt{\sum \alpha_{ir}^2 q_i q_r}. \qquad (52)$$

Thus if we minimise the average value (for all k species) of $\sum_{r=1}^{k} \alpha_{ir}^2 q_i q_r$, we will minimise the frequency of gross fluctuations in population size which might lead to extinction, and hence will maximise what I have called persistence.

Some restraint must be placed on the scale and nature of the ecosystem for which (52) is to be minimised; otherwise we could minimise (52) by putting all the $\alpha_{ir} = 0$. Leigh minimises $\Sigma\alpha_{ir}^2 q_i q_r$ under the restraint that the total productivity P of the ecosystem should be constant. How is P to be measured? If we consider the ecosystem divided into primary producers and others, then the primary producers are those for which ϵ_r in (46) is positive, and the average value of the total primary production is $\Sigma\epsilon_r q_r$, summed for all primary producers. For an ecosystem with only two trophic levels, this sum is equal to

$$\tfrac{1}{2}\sum_{rs} |\alpha_{rs}| q_r q_s$$

when the ecosystem is at equilibrium. A somewhat more complex expression is needed if there are more than two trophic levels, but for simplicity Leigh defines

$$P = \tfrac{1}{2}\sum_{rs} |\alpha_{rs}| q_r q_s.$$

It is easy to see that if P is constant, then $\Sigma\alpha_{ir}^2 q_i q_r$ will be minimised if all the α_{ir} are equal. Hence the persistence of an ecosystem is increased if there are a large number of weak connections between the component species rather than a small number of strong ones.

Given that all the α_{ir} are equal, and provided that the number of species is not too small, it can be shown that the frequency (52) is proportional to

$$\frac{P}{B} \sqrt{\left[\frac{1}{k} \exp\left(-q_i a^2 / 2\theta\right)\right]}, \tag{53}$$

where B is the total biomass of the system, $\sum_i N_i$.

Thus if the connectedness of the food web is sufficiently extensive, the persistence of the system is increased if there is a fall in productivity relative to biomass.

So much for an outline of the theory. Unfortunately it is doubtful whether the theory can help towards an understanding of the persistence of ecosystems. Thus the basic equations (46) on which the whole theory rests describe a conservative system, in which the violence with which species fluctuate in numbers depends entirely on the initial conditions. This is necessary, since the methods of statistical mechanics cannot be used on non-conservative systems. But at the same time it means that the approach can tell us nothing about what determines the violence of fluctuations in numbers.

The difficulty becomes more apparent if we consider two predictions of the theory. Consider first Kernet's derivation of a species abundance distribution, which might seem to have some observational support. In fact Preston (1962) has shown that the 'log normal' distribution of species abundance is a better fit to observational data than that chosen by Fisher and predicted by Kerner. But this is a minor objection. The real objection to (51) is that it is not a predicted distribution of N_r, the actual species abundances (or rather biomasses) but of $n_r = N_r/q_r$. Thus Kerner's theory only predicts an appropriate distribution when all species are assumed to have the same equilibrium density q_r, and to be fluctuating with large amplitude (θ large). Thus the predicted distribution reflects the departure of species abundances from their average values. It is an implication of this that if two samples were taken from the same ecosystem at two times far enough apart for a number of cycles to have intervened, both samples would show the same species abundance distribution, but the common species in one sample would be rare in the other, and *vice versa*. If, as is likely to be the case in practice, the same species are found to remain common in successive samples, then the distribution has little to do with fluctuations, and reflects instead differences between the equilibrium value q_r, about which the theory can say nothing.

Similar difficulties arise in interpreting Leigh's analysis of the persistence of ecosystems, and his demonstration that in a system of constant productivity, persistence is maximised if all the interactions α_{ir} are equal. Thus the violence of fluctuations in numbers depends on the initial conditions. All that the inter-connectedness of the food web can do is to alter the form of these fluctuations. The point is illustrated in figure 38. A graph of N_r against time for any species is likely to resemble A if the food web has a small number of strong connections, and to resemble B if there is a

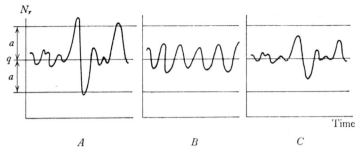

Figure 38. The number N_r of a species against time for the Kerner–Leigh model. *A*, for a food web with a small number of strong interactions; *B*, for a web with a large number of weak interactions; *C*, as *A*, but with initial conditions closer to the equilibrium. These are not computed curves, but do illustrate qualitatively the differences to be expected.

larger number of weak connections. There is nothing in the theory to show that an increase in connectedness changes the graph from *A* to *C*, since this difference depends solely on initial conditions. Thus curves *A* and *B* are at the same 'temperature' θ and have the same average (amplitude)^2 of oscillation, but *A* crosses the line $N = (q+a)$ more often because the amplitude is more variable. This is a difference which could exist between two species in the same ecosystem (*A* having fewer first-order interactions than *B*), or between all species in one ecosystem and all the species in another system weakly coupled to the first and therefore at the same temperature. But the difference between *A* and *C* is a difference of temperature, not dependent on connectivity, but on initial conditions, or on environmental fluctuations. Nevertheless, the difference between patterns *A* and *B* appears to be a real prediction of the theory, and it may well be one which holds also for non-conservative systems.

9 COMPLEXITY AT A SINGLE TROPHIC LEVEL

A. Conditions for coexistence of competing species

It is a well-known conclusion from the Lotka–Volterra model of two-species competition that stable coexistence requires that the competition parameters must satisfy two inequalities (see Chapter 5). Strobeck (1973) has shown that if the corresponding equations for competition between n species are considered, then stable coexistence of all n species requires that $2(n-1)$ inequalities must be satisfied. This conclusion should be borne in mind when considering the stability of n-species ecosystems with predator–prey and commensal interactions as well as competitive ones. Strobeck's finding is another argument in favour of the view that large ecosystems with arbitrary interactions are unlikely to be stable.

B. How many limiting resources?

It was shown by MacArthur and Levins (1964), and independently by Rescigno and Richards (1965) that there cannot be more species in a habitat than there are independent resources which limit them. An earlier statement of this idea was given by Williamson (1957). The difficulty in applying this theorem lies in deciding how many independent resources there in fact are. I will first give MacArthur and Levins' version, since although less rigorous it is easier to interpret ecologically. I will then give some apparent counter examples, and then in the light of these will attempt to state more precisely what is meant by two resources being independent. Finally in this section I will describe an extension of the argument due to Levin (1970).

Consider first the case of two resources, densities R_1 and R_2. It is required to prove that not more than two species can be limited solely by these resources. If species A, density y_A, is

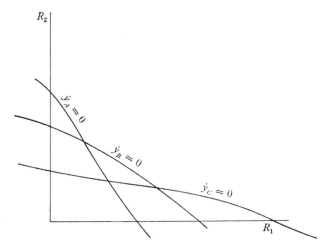

Figure 39. Three species with two limiting resources.

limited only by these resources, then there is a line in the $(R_1 R_2)$ plane for which $\dot{y}_A = 0$ (see figure 39). Similarly for species B, C, etc., there will be lines $\dot{y}_B = 0$, $\dot{y}_C = 0$ and so on.

If two species A and B are to coexist, then there must be some point in the plane where $\dot{y}_A = 0$ and $\dot{y}_B = 0$; that is, the two lines must intersect in a point. It is, however, infinitely improbable (a phrase used by MacArthur and Levins) that three lines should intersect in a point, and hence three species cannot coexist if limited by only two resources. A word must be said in justification of this argument. There is nothing in mathematics to prevent three lines meeting in a point. But in the ecological situation represented by this model, any change in a species (for example, a change in a gene frequency) would alter the position of the corresponding line. Hence the argument that it is infinitely improbable that three lines should meet in a point is sound, unless there is some stabilising process causing them to do so. But to assume such a stabilising process would be to assume that selection acts directly so as to stabilise ecosystems, which is not the case.

If the argument is accepted for two resources, no difficulty exists in extending it to n resources, and stating that not more than n species can coexist in a habitat with n limiting resources. Note that

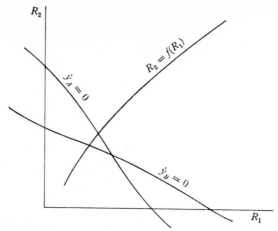

Figure 40. Two species dependent on two resources whose densities are functionally related.

the argument places an upper limit on the number of species; it does not prove that if there are n resources there need be n species.

Consider now two apparent counter examples:

(i) Two predators are limited by a single prey species, one feeding on the larvae and the other on the adult stage.

(ii) Two herbivores are limited by a single plant species, one feeding on the roots and the other on the leaves.

In both these cases, it is easy to formulate plausible models giving stable coexistence of the two species (for further details, see Haigh and Maynard Smith, 1972). Thus if the theorem is to be maintained, it is necessary to regard the larvae and adults of the same insect species, or the roots and leaves of the same plant species, as independent resources. It is clear that to be independent for the purposes of the theorem, two resources need not be statistically independent.

Suppose, however, that the densities of two resources are functionally related, so that $R_2 = f(R_1)$. Then it is clear from figure 40 that for an equilibrium between species A and B to exist, three curves must meet in a point, and we have already argued that this is infinitely improbable.

The simplest reason why the densities of two resources may be functionally related is that they are 'fine-grained' in Levins' (1968) sense. If a species always takes two resources in the proportions in which they exist in the habitat (for example, a filter feeder unable to discriminate between two food organisms), then the resources are not independent and should be regarded as a single resource. Note that two food items may constitute two independent resources or a single resource, depending on the nature of the predator. Thus the liver and intestines of a mouse are a single resource to an owl, but two independent resources to parasitic organisms.

Levin (1972) has extended this theorem. He considers the fate of n species, densities $x_1, x_2, \ldots, x_n$, which obey the equations

$$\dot{x}_i = x_i f_i(x_1, \ldots, x_n, y_1, \ldots, y_m). \tag{54}$$

In these equations $y_1, \ldots, y_m$ represent variables, such as climate or invading species, which are not functions of the x's, but which may be functions of t. Apart from these variables, the only 'resources' appearing as limiting factors on the right-hand side of the equations are the species densities themselves. This does not mean that a fixed resource such as space cannot act as a limiting factor, because the space per individual of species i is a function of x_i, and would enter in that form into the equation for x_i.

Levin then introduces new variables $z_1, \ldots, z_p$, where $p < n+m$. Each z is a function of the x's and y's, and the z's are chosen so as to be a 'minimum independent set', such that each f_i can be expressed as a function of the z's. Levin then proves that no stationary equilibrium or continuous cycle is possible unless $p \geqslant n$. Thus the z's are the 'limiting factors' or 'resources', and there must be at least as many independent resources as there are species to be regulated.

The variables z are introduced in order to cover the point which was discussed earlier that two resources must not be functionally dependent on one another. Thus suppose that whenever x_1 or x_2 appeared on the right-hand side of an equation, they appeared in the form $(x_1 + 4x_2)$; then it would be appropriate to replace $(x_1 + 4x_2)$ by z_1, a single limiting factor.

This theorem can be simplified if it is appreciated that the variables $y_1, \ldots, y_m$ are irrelevant. Thus consider the equations

$$\dot{x}_1 = x_1 f_1(x_1, y), \tag{55}$$

$$\dot{x}_2 = x_2 f_2(x_1, y), \tag{56}$$

where y is not a function of x_1 or x_2, but may vary with time. There is nothing in Levin's formulation of the theorem to show that no equilibrium can exist, since there are two limiting factors x_1 and y.

Equation (55) can be rewritten

$$\dot{x}_1 = F_1(x_1, t),$$

which, given initial conditions, will have a solution

$$x_1 = g(t).$$

Thus equation (56) becomes

$$\dot{x}_2 = x_2 F_2[g(t), t],$$

that is, $\dot{x}_2 = x_2 h(t)$.

Consider the equation $\dot{x}_2 = x_2[h(t) + \epsilon]$, where ϵ is small but non-zero. The equation has the solution

$$x_2 = K \exp(\epsilon t) \exp\left[\int h(t)\, dt\right]. \tag{57}$$

In our case $\epsilon = 0$, so, for equilibrium, $K \exp\left[\int h(t)\, dt\right]$ must be bounded away from zero or infinity. Hence the introduction of a small non-zero ϵ, which could arise from a small change in the biology of either species, will cause the right-hand side of equation (57) either to diverge to infinity or decrease to zero according as $\epsilon > 0$ or $\epsilon < 0$. Hence the original system (55), (56) is not stable.

The argument can readily be extended to n species (Haigh and Maynard Smith, 1972). This does not invalidate the rest of Levin's argument. Thus given n equations of the form

$$\dot{x}_i = x_r f_i(x_1, \ldots, x_n, t), \tag{58}$$

Levin shows that for there to be a stable equilibrium requires:

(i) There is no subset of r self-contained equations in which fewer than r densities appear as limiting factors. As a special case, all n densities must appear in at least one limiting function f_i.

(ii) It must not be possible to replace the variables $x_1, \ldots, x_n$ by variables $z_1, \ldots, z_p$, where $p < n$, such that $f_1, \ldots, f_n$ can be expressed as functions of $z_1, \ldots, z_p, t$.

Condition (i) says something about the minimum complexity of stable ecosystems, including the fact that in any ecosystem with n species, the density of each species must influence either its own rate of increase or the rate of increase of some other species in the system.

Condition (ii) appears to be a very general form of 'Gause's principle'. This principle has been variously formulated;[*] a typical formulation is that no two species which have identical ecological requirements can long coexist in a habitat. Condition (ii) states that if two species are identical (or if one member of one species is identical to r of the other) in their role as limiting factors, no ecosystem containing both can be stable.

---

[*] But never by Gause; for an account of the origin and significance of Gause's principle, see Hardin (1960).

10 COMPLEXITY WITH SEVERAL TROPHIC LEVELS

A. Food chains with more than one step

It was shown at the beginning of the previous chapter that as the number of competing species on a single trophic level increases, there is a corresponding increase in the number of inequalities which must be satisfied if there is to be a stable equilibrium. There is no corresponding increase in the number of stability conditions as the length of a food chain is increased. Thus consider a prey species, density x, and three predator species, densities y, z and w, each feeding on the one below it in the food chain. Then if we assume that the prey species is resource-limited, and the predators limited only by their prey, we have:

$$\left. \begin{aligned} \dot{x} &= x(a-bx-c_1y), \\ \dot{y} &= y(-e+c_1'x-c_2z), \\ \dot{z} &= z(-f+c_2'y-c_3w), \\ \dot{w} &= w(-g+c_3'z), \end{aligned} \right\} \tag{59}$$

where all the coefficients are positive. It is easy to show that there is an equilibrium point with x, y, z and w positive provided that

$$\frac{a}{b} > \frac{e}{c_1'} + \frac{fc_1}{bc_2'} + \frac{gc_2}{c_1'c_3'}. \tag{60}$$

This corresponds to the condition $a/b > e/c$ which must be satisfied for the coexistence of a single predator and its prey. Although the equilibrium density a/b of the prey on its own has to be greater if it is to support three predators than if it had to support only the first in the chain, there is still only a single inequality to be satisfied. This conclusion holds for a food chain of any length. I have not investigated the stability of the equilibria, but it seems likely that they will be stable.

Hence the lengthening of food chains will not by itself cause instability.

B. Ecosystems with random interactions

Two methods of approach have been made to the study of complexity in ecosystems with predator–prey as well as competitive interactions. The first, considered in this section, is to analyse the effects on the stability of an ecosystem of random variations in the sign and magnitude of the interactions between species. This approach has been outlined in an important paper by May (1971b).

May first considers a general extension of the Volterra equations without resource limitation (p. 22). He considers an ecosystem with n predator species, densities $P_1, P_2, \ldots, P_j, \ldots, P_n$, and n prey species, densities $H_1, H_2, \ldots, H_j, \ldots, H_n$; here H can be thought of as standing for herbivore. The equations analogous to Volterra's equations are

$$
\left.
\begin{aligned}
\mathrm{d}H_j/\mathrm{d}t &= H_j\left[a_j - \sum_{k=1}^{n} \alpha_{jk} P_k\right], \\
\mathrm{d}P_j/\mathrm{d}t &= P_j\left[-b_j + \sum_{k=1}^{n} \beta_{jk} H_k\right],
\end{aligned}
\right\}
\tag{61}
$$

where $j = 1, 2, \ldots, n$.

These equations differ from those considered by Kerner (p. 92) in that it is *not* assumed that $\alpha_{jk} = \beta_{kj}$; the only restriction placed on the coefficients is that they should give positive finite values for all densities at equilibrium. May (1971b) shows that *either* the behaviour is purely oscillatory, as is always so for the two species case, *or* the system is unstable. The conclusion is clear: it is not a necessary consequence of increasing the complexity of an ecosystem that it becomes more stable, and it is at least possible that it will become less stable.

May then goes on to consider an ecosystem of m species, densities $N_1, N_2, \ldots, N_m$, obeying the rather general equations

$$
\mathrm{d}N_j/\mathrm{d}t = F_j(N_1, N_2, \ldots, N_m),
\tag{62}
$$

with $j = 1, 2, \ldots, m$.

Here the m functions F_j are arbitrary non-linear functions of the population densities. The equilibrium points are given by the solutions of the set of m equations

$$
F_j(\hat{N}_1, \hat{N}_2, \ldots, \hat{N}_m) = 0.
\tag{63}
$$

It is an implicit restraint on the functions F_j that these equations have at least one solution for which the values $\hat{N}_1, \hat{N}_2, \ldots, \hat{N}_m$ are positive. The stability of the equilibrium for small displacements is analysed by putting $N_j = \hat{N}_j(1 + n_j)$, when the initially small quantities n_j measure the disturbance from the equilibrium population. Equations (62) are then reduced to the linear differential equations

$$\mathrm{d}n_j/\mathrm{d}t = \sum_{k=1}^{m} a_{jk} n_k, \tag{64}$$

where a_{jk} is given by
$$a_{jk} = \frac{\hat{N}_k}{\hat{N}_j} \frac{\partial F_j}{\partial N_k},$$

the partial differential being evaluated at the equilibrium.

The stability of the system then depends on the matrix of elements a_{jk}.

Suppose now that each species by itself has a stable equilibrium density, which it approaches at a constant rate. This amounts to assuming that
$$a_{jj} = -1, \quad \text{for all } j. \tag{65}$$

It is then assumed that all other a_{jk} (with $j \neq k$) are chosen from a random distribution with mean 0 and variance S^2. We can think of S as measuring the average strength of the interaction between species. The interactions can be of either sign, and considering pairwise interactions, competitive and commensal interactions are equally common, and taken together are as common as predator–prey interactions. May then shows that if $m \gg 1$, then if
$$S < (2m)^{-\frac{1}{2}}, \tag{66}$$

the system is almost certainly stable, and if

$$S > (2m)^{-\frac{1}{2}}, \tag{67}$$
the system is unstable.

May then goes on to conjecture that if the a_{jk} were drawn from the above statistical distribution, with probability C, and were put equal to zero with probability $1 - C$, then the stability criterion corresponding to (66) would become

$$S < (2mC)^{-\frac{1}{2}}. \tag{68}$$

The quantity C represents the probability that any two species interact directly, and is referred to as the 'connectance' of the web. This analytical result agrees well with computer simulations by Gardner and Ashby (1970), for $m = 4, 7$ and 10.

How is this result to be interpreted? It states that if a number of species each has a stable equilibrium when isolated, then as the number of species m or the connectance C increases, the average strength of interactions S between species which is compatible with stability decreases. This again leads to the conclusion that if large and richly connected natural ecosystems are stable, this can only be because the connections are highly non-random.

If one supposes that the stability of real ecosystems arises from 'species exclusion' as described on p. 89, one can ask further questions of May's model. May has shown that an ecosystem of m species with randomly chosen interactions is unlikely to be stable if m is large. But we can ask, if a total of m species are introduced into a community, what is the expected number r (where $r \leqslant m$) of species which will remain in a stable system when the exclusion of species has been completed. The answer would depend on whether the species were introduced one by one or simultaneously, and on other assumptions. But if the answer turns out to be that if m is large then r/m is not greatly less than unity, and if the 'species exclusion' model is appropriate to continental as well as to island systems, then the existence of complex ecosystems can be explained without recourse to 'genetic feedback'. But these are big 'ifs'.

C. Systematically connected webs

An alternative approach is to consider an ecosystem in which the connectivity is altered in a systematic way while maintaining other features of the system unaltered. As an example, consider the ecosystem illustrated in figure 41, composed of two prey species, two intermediate predators and two top predators. It is supposed that the two prey species are resource-limited. The connectivity of the system can be altered by increasing or decreasing the importance of the arrows C (measuring competition between

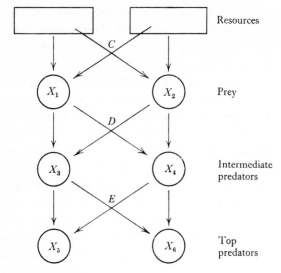

Figure 41. Six-species ecosystem. The arrows represent flow of material.

the prey), D (measuring competition between the intermediate predators), or E (measuring competition between the top predators).

The system can be described by the equations:

$$
\begin{aligned}
X_1' &= X_1(a_{10} - a_{11}X_1 - a_{12}X_2 - a_{13}X_3 - a_{14}X_4 &&), \\
X_2' &= X_2(a_{20} - a_{21}X_1 - a_{22}X_2 - a_{23}X_3 - a_{24}X_4 &&), \\
X_3' &= X_3(a_{30} + a_{31}X_1 + a_{32}X_2 && - a_{35}X_5 - a_{36}X_6), \\
X_4' &= X_4(a_{40} + a_{41}X_1 + a_{42}X_2 && - a_{45}X_5 - a_{46}X_6), \\
X_5' &= X_5(a_{50} && + a_{53}X_3 + a_{54}X_4 &&), \\
X_6' &= X (a_{60} && + a_{63}X_3 + a_{64}X_4 &&).
\end{aligned}
$$

$$(69)$$

In these equations, all the a_{ij} are positive. The coefficients a_{10} and a_{20} are greater than one, and a_{30}, a_{40}, a_{50}, a_{60} are less than one.

Two underlying assumptions should be noted:

(i) The equations are in finite difference form, which is equivalent to assuming discrete generations. However, they also give an approximate description of an ecosystem with continuous reproduction but with delayed regulation.

(ii) Only the two prey species have self-limiting terms. Thus it is assumed that the prey species are resource-limited, and the other species are limited only by their food.

The behaviour of the system will depend on the relative importance of the delays introduced by discrete breeding seasons, and the damping due to self-limiting terms.

There are 26 arbitrary constants in equations (69). It is required to reduce these to three parameters, c, d, and e, measuring the connectivity of the web at the three levels. To do this, we proceed as follows:

(i) We let $a_{ij} = a_{ji}$ throughout, except for the coefficients a_{12} and a_{21}. This implies that $X_1, \ldots, X_6$ refer to 'equivalent biomass' and not to numbers of individuals. Thus if, for example, n grams of prey are converted into 1 gram of intermediate predator, and if X_1 and X_2 represent the actual biomass of the prey, then X_3 and X_4 represent n times the actual biomass of the predators.

(ii) The equilibrium values $\hat{X}_1, \ldots, \hat{X}_6$ of each species are kept constant and equal to unity. Since they represent equivalent biomasses rather than actual ones, the assumption of equality of biomass between levels is not unrealistic. The choice of unity is merely a scaling factor.

(iii) We choose the intrinsic rates of increase $a_{10}, \ldots, a_{60}$ as follows. When there are no interactions between the two chains, $X_1 - X_3 - X_5$ and $X_2 - X_4 - X_6$, the values $a_{10}, \ldots, a_{60}$ are such as to ensure that each chain oscillates about its equilibrium value with uniform amplitude. To avoid resonance effects, the periods of oscillation of the isolated chains are different, of six and eight generations respectively.

Assumption (i) provides eight equations, assumption (ii) provides six equations and assumption (iii) a further six equations connecting the twenty-six coefficients. This leaves six equations to describe the strength of the connections between the two chains. These are expressed in terms of three parameters, as follows:

$$\left.\begin{aligned}
a_{12} &= ca_{11}; & a_{21} &= ca_{22}; \\
a_{14} &= da_{13}; & a_{23} &= da_{24}; \\
a_{45} &= ea_{35}; & a_{36} &= ea_{46}.
\end{aligned}\right\} \tag{70}$$

These equations should be interpreted as follows. If $c = 0$, then there is no competition between the prey species. If $c = 1$, there is maximum competition between the prey species, each inhibiting its competitor and itself equally. If $c > 1$, then the equilibrium between the prey species would be unstable in the absence of any predation. The parameters d and e have similar meanings for competition between intermediate and top predators respectively.

With assumptions (i) to (iii) and equations (70) it is possible to calculate the coefficients a_{ij} as functions of c, d, and e. If we now write $X_i = \hat{X}_i(1 + x_i)$ for all i, and consider only small displacements from the equilibrium, equations (69) become

$$
\begin{pmatrix} x_1' \\ x_2' \\ x_3' \\ x_4' \\ x_5' \\ x_6' \end{pmatrix} = \begin{pmatrix} (1 - a_{11}) & -a_{12} & -a_{13} & -a_{14} & 0 & 0 \\ -a_{21} & (1 - a_{22}) & -a_{23} & -a_{24} & 0 & 0 \\ +a_{31} & +a_{32} & +1 & 0 & -a_{35} & -a_{36} \\ +a_{41} & +a_{42} & 0 & +1 & -a_{45} & -a_{46} \\ 0 & 0 & +a_{53} & +a_{54} & +1 & 0 \\ 0 & 0 & +a_{63} & +a_{64} & 0 & +1 \end{pmatrix} \begin{pmatrix} x_1 \\ x_2 \\ x_3 \\ x_4 \\ x_5 \\ x_6 \end{pmatrix}.
$$

$$(71)$$

These equations can readily be iterated on a computer, for a sufficient number of generations to investigate their stability. This has been done for a range of values of c, d, and e. The results are shown graphically in figure 42. In interpreting this figure, it should be remembered that when in isolation ($c = d = e = 0$) the chains oscillate with uniform amplitude. Thus an arrow pointing upwards indicates that connectivity has caused instability, and an arrow pointing downwards that it has caused stability. It is clear that connectivity can either stabilise or destabilise the system, depending on the level at which it occurs. In general, competition between the prey species is destabilising, whereas competition between the top predators (*i.e.* generalist rather than specialist predators) is stabilising. Competition at the intermediate level has inconsistent effects.

There is one other feature of the behaviour of this system which is not apparent from the figure. If the connectivity is such as to

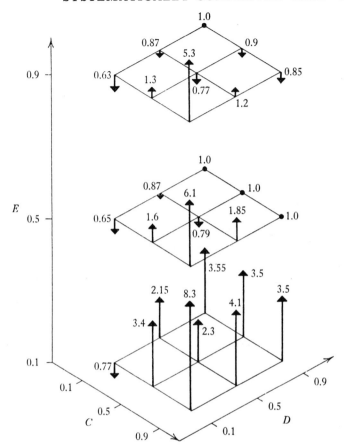

Figure 42. Behaviour of six-species ecosystem, for various values of *c*, *d* and *e*. In all cases the behaviour which dominates as $t \to \infty$ is an oscillation. The numbers against the arrows represent the ratio of the amplitude of one cycle to that of the preceding cycle; the lengths of the arrows are proportional to the logarithms of this ratio. Thus upwardly directed arrows represent divergent oscillations.

stabilise the system, then the two chains oscillate in phase, with a period of approximately seven generations, intermediate between the periods of the two isolated chains. If the connectivity destabilises the system, the two chains oscillate out of phase, one prey species increasing while the other decreases, with a period of between ten and twenty generations.

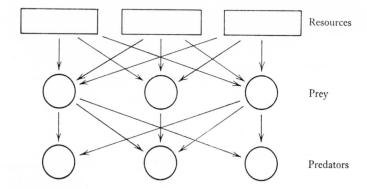

Figure 43. An alternative six-species ecosystem.

The main conclusions to be drawn from this model are:

(i) Competition at the prey level of an ecosystem is destabilising.

(ii) Competition between predators (*i.e.* generalist rather than specialist predators) is stabilising.

(iii) When connectivity destabilises a system, its components oscillate out of phase, with a period longer than their period in isolation.

It is important to know whether these conclusions depend on the particular structure of the ecosystem in figure 41, or on the various additional assumptions that have been made. It does not seem that they do. Thus exactly the same conclusions emerged from a similar analysis of the system shown in figure 43. Investigation has also shown that the conclusions are robust to changes in the natural period of the isolated chains, and in the equilibrium values of the biomasses of the various species.

D.   The relevance of models to field data

What is the relevance of the models discussed in this and the last chapter to the view that stability is a consequence of complexity? As a start, it is worth reviewing the attempt by Watt (1964, 1965) to decide whether, in nature, complexity of the food web makes for stability or instability. Watt analysed very extensive data from the Canadian Forest Insect Survey on the population numbers of

Macrolepidopteran species, in terms of the food plants upon which they feed, and of their taxonomic relationships. He summarised his conclusions by saying that the stability at any herbivore or carnivore trophic level:

(i) increases with the number of competitor species at that level,

(ii) decreases with the number of competitor species that feed on it, and

(iii) decreases with the proportion of the environment containing useful food.

'Stability' refers to a single species, and is greatest when the numbers of a species vary least (on a logarithmic scale) from year to year. Watt's evidence for statement (i) is that those species feeding on food plants also taken by many other species fluctuate in numbers less than those feeding on food plants taken by few other species. His evidence for statement (iii) is twofold: first, species with a large number of food plants fluctuate more than those with few food plants; and second, the major exception to this rule, the spruce budworm, which confines itself to two food species but is nevertheless highly unstable, is unstable only in those areas where its food trees form the major part of the environment. Watt's statement (ii) is not supported by his own analysis, and is introduced to account for Zwolfer's (1963) finding that lepidoptera attacked by several parasites are more likely to 'escape' and fluctuate out of control than those attacked by only one parasite.

How can these conclusions be explained? Watt's conclusion (iii) is fairly easily understood. A herbivore is likely to fluctuate with large amplitude if its intrinsic rate of increase is high, and this will be so if its food is readily available and densely distributed, so that few individuals are lost during dispersal. It is more difficult to understand conclusions (i) and (ii). At first sight these conclusions seem to be the opposite of those which emerge from the six-species ecosystems just considered. However, the comparison is clearly not valid. Watt's conclusions concern the relative amplitudes of fluctuation of different species within the same ecosystem; the model compares the stability of different systems. In particular conclusion (ii) suggests that if the species on one trophic level are incapable of rapid increase and hence stable, then the species on the trophic level below may be unstable.

It follows that Watt's analysis is important for an understanding of the relative stability of the species within an ecosystem. The models discussed in this chapter, which compare the stability of different ecosystems, are only indirectly relevant to an understanding of Watt's conclusions.

There is a more specific reason why the models considered here, and in particular May's model, may not be strictly relevant to ecological findings based on field data. The arguments adduced by Elton (1958) and briefly summarised at the beginning of Chapter 7 are all of one kind; they state that divergent predator–prey oscillations (using this term to include all $+ -$ interactions) are more often observed in ecosystems with few species. Now May's model does not in any case permit the occurrence of a divergent predator–prey oscillation. Thus if N_j and N_k are the numbers of a prey and predator species respectively, the equations for the departure of these numbers from the equilibrium are

$$\dot{n}_j = -n_j - a_{jk}n_k,$$
$$\dot{n}_k = a_{kj}n_j - n_k,$$

where a_{jk} and a_{kj} are both positive. It is easy to show that these equations cannot lead to a divergent oscillation. In essence, the equations assume that all species, including predators, are self-limited, and they ignore time delays. Hence most of the instabilities that arise will be of a 'competitive exclusion' type, and any instabilities of an oscillatory nature will arise from multi-species interactions rather than from simple predator–prey interactions.

It follows that May's model does not say anything one way or the other about the suggestion that multi-species ecosystems are more stable because they make predator–prey oscillations less likely. Yet such a suggestion is supported by Watt's (1965) observation that a phytophagous insect is more likely to fluctuate in numbers if its food plant or plants form a large part of the environment; similar observations have been made by other workers (e.g. Satchell, 1962, for defoliation of oaks by *Tortrix*). It is also in line with an interesting suggestion by Connell (1970) concerning the large numbers of competing species in tropical ecosystems. He argues that in a topical rain forest no single species of tree can become abundant, because if it does so it is attacked by specialist predators and diseases. This amounts to saying that each tree species is kept to a low equilibrium density $\hat{x}$, well below the carrying capacity x_e which it could reach in the absence of predation or disease. At first sight this might suggest that oscillatory instabilities would be common. Thus it was argued (p. 28) that predator–prey oscillations are most likely when $\hat{x} \ll x_e$. However, in this case no tree species can in the short term increase above the equilibrium density $\hat{x}$ because of competition from other tree species, and hence oscillations cannot develop. It may be that in the long run, if all predators were removed, one or a few tree species would competitively exclude all the others, but this process would be very slow compared to the predator life cycle.

One can therefore envisage a stable ecosystem in which each tree species is relatively rare and in equilibrium with its specific predators and diseases. These equilibria are stable because competition prevents rapid increase of any tree species. At the same time, competitive exclusion of tree species is prevented by specific predation. Such a system could be stable, although a system composed of only a few tree species and their predators might not be.

It is not clear why similar complexity is not found in temperate and arctic systems. Connell suggests that 'predators' are more susceptible to climatic extremes that their 'prey', and hence are only able to prevent prey species from reaching high densities in tropical conditions. This may be correct, but it is not clear why such a difference in susceptibility should exist.

To summarise, mathematical models such as those of May (1971b) and Strobeck (1973) make it clear that stability is not a necessary or even a likely consequence of increased complexity. There may nevertheless be specific cases in which an increase in the number of species in an ecosystem may increase its stability. The observational evidence suggests that divergent predator–prey oscillations are more likely in ecosystems with few species; there are no theoretical grounds for doubting this conclusion.

11 COEVOLUTION

A. Genetic feedback

If an increase in the number of species and of inter-species inter-actions does not in itself lead to stability, the persistence of eco-systems implies that the interactions are highly non-random; that is, that they are consequences of some type of selection. There are two ways in which we can imagine such selection acting. First, we can suppose that the genetic properties of species are fixed (or are evolving in ways which do not regularly lead to stability). The actual groups of species which we find in any region are a subset of those which have in the past reached that area. They are necessarily a subset capable of persistence, because if they were not we would not find them. They must also be a subset capable of excluding most new invaders, because if they were not, the species occupying a given region would change more often and more drastically than seems to be the case. According to this explanation of persistence, what is required is that of the N species reaching an area, some subset r should form a persistent ecosystem capable of resisting the introduction of most of the remaining $(N-r)$ species. This is much more probable than that all N should persist; it is the latter which arguments such as those of Strobeck (1973) and May (1971 b) show to be unlikely for large N.

Alternatively, selection could account for persistence if it were the case that when two species coexist for some time, the effects of selection upon them were to alter them so as to make stable coexistence more likely. The main proponent of this view, under the name 'genetic feedback', has been Pimentel (1968; also Pimentel and Soans, 1970). Two different mechanisms have been suggested or at least implied for genetic feedback; I will describe the mechanisms first and then discuss the evidence for them.

The first suggestion is that as the relative abundance of two species alters, the intensity of selection acting on them will alter so as to make for stability. Thus consider two competing species, A and B. If A is abundant and B rare, most of B's interactions will be with A, and so selection will act so as to make B more efficient in interspecific competition; for the same reason it will make A more efficient in intraspecific competition. The effect of these changes will be to increase the abundance of B relative to A. Thus selection will tend to alter both species so that the rare species becomes more abundant.

A similar argument applies to the coexistence of a predator and its prey, though with less force. Thus consider the model of a vertebrate predator and its prey in Chapter 4E; this is chosen because the searching success of the predator appears explicitly in the model, but the argument applies to any predator–prey interaction. It was shown for that model that persistence requires that the effective search area α should lie between certain limits; if α is too small, the predator goes extinct, and if α is too large both species go extinct. Selection acting on the predator will tend to increase α, and on the prey to decrease it. If the predator is rare and consequently a relatively unimportant cause of death in the prey, the intensity of selection on the prey for improved defence or escape mechanisms will be low. Since defence in general costs energy, in the absence of intense selection it is likely to diminish in efficiency; if so α will increase, and with it the numbers of predators. Unfortunately it is less clear that the reverse process will operate. A common predator which is an important cause of death to its prey will still be under intense selection for increased α.

A second mechanism of genetic feedback has been implied rather than stated explicitly by Pimentel. This is that selection favouring the survival of whole populations – that is, group selection – has favoured ecological stability. Thus in such statements as 'if an animal feeds upon the "capital" (energy necessary for growth, maintenance and reproduction) it will eventually destroy both its food resource and itself' (Pimentel and Soans, 1970), the word 'animal' can only refer to a population of animals. The same idea apparently lies behind the argument that a predator will evolve so as to subsist on 'interest' rather than 'capital'.

5

This particular concept does not seem helpful, since if there is a stable equilibrium between a predator and its prey, it is by definition true that the predator will be subsisting on interest; we are therefore left with the original problem of explaining the existence of a stable equilibrium. In so far as the concept of genetic feedback depends on mechanisms of group selection, it is the same as the concepts worked out in great detail by Wynne-Edwards (1962).

The best evidence for genetic feedback acting on competing species comes from studies of competition between the housefly, *Musca domestica*, and a blowfly, *Phaenicia sericata* (Pimentel *et al.*, 1965). When houseflies and blowflies derived from wild populations competed in a small cage, the result was indeterminate, sometimes one and sometimes the other surviving. In an experiment in which the two species competed in a large cage, the housefly initially increased in numbers and the blowfly only just maintained itself in low numbers. After 55 weeks (approximately 2 weeks per generation) the blowfly began an explosive increase, and the housefly generally became extinct after 65 weeks. This suggests that the initially rarer species evolved so as to increase in competitive ability, as predicted by the theory.

To test this, competition experiments in small cages were carried out between 'experimental' flies modified by competition in the large cage and 'wild' flies unmodified by competition; the former were taken from the large cage after 38 weeks. The results are shown in table 3. In general, they conform to prediction. The blowflies seem to have increased in competitive ability, and there is no evidence that the houseflies have done so.

Pimentel and Al-Hafidh (1965) made a similar study of genetic feedback in a host insect, again *Musca domestica*, and a parasitic wasp, *Nasonia vitripennis*, which lays eggs in the host pupae. There is evidence that the flies became more resistant to the parasite, but evidence for changes in the parasite is less convincing. A very different example of genetic feedback in a predator–prey system (Horne, 1970) concerns *Escherichia coli* and T-even bacteriophages in chemostat culture. This system initially fluctuates with enormous amplitude, but Horne found that repeatable

TABLE 3. *Coevolution of houseflies and blowflies. Results of competition in small cages between flies which had been modified by approximately 20 generations of coexistence (coevolved), and unmodified flies (Pimentel et al., 1965). During the period of coevolution, the blowflies were relatively rare and the houseflies common*

Origin of flies		Number of cages		
Houseflies	Blowflies	Total	Won by housefly	Won by blowfly
Unmodified	Unmodified	9	6	3
Unmodified	Coevolved	5	0	5
Coevolved	Unmodified	5	3	2
Coevolved	Coevolved	5	0	5

changes occur in both host and parasite leading to a relatively stable equilibrium with infective phage present in very low frequency. The nature of the genetic changes is not known; two possibilities are, first, that the bacteria have evolved resistance and that the phage are surviving in the back-mutants to susceptibility; and second, that the phage have become temperate.

There is therefore some experimental support for the concept of genetic feedback. The examples given seem to be caused by the first of the two suggested mechanisms; that is, by changes in selection intensity with relative abundance.

B. Specialist or generalist?

Whether a particular species will be a specialist, depending on one or a few resources, or a generalist, depending on many resources, will depend on the operation of natural selection. The purpose of the model developed in this section is to determine what circumstances will lead to one or other pattern of behaviour. I consider first a single species, and later analyse the effects of competition from other species.

Let x = density of 'predator' species,

R_j = density of the jth resource, assumed constant.

p_j = probability that, when an encounter occurs with resource j, an 'attack' takes place.

D_j = time spent in an 'attack' on resource j, until the predator is ready to start hunting again.

W_j = expected gain (measured in units of x) from an attack on resource j, allowing for the possibility of not capturing the resource.

n_j = number of attacks on resource j, per predator per unit time.

Natural selection can alter p_j, D_j, and W_j. Selection will tend to maximise W/D for any resource for which p is non-zero, but it will usually be the case that an increase in W/D for one resource will entail a decrease for other resources. Values of p can be altered by selection, and may alter by learning without genetic change; in general there is no reason why the value of p for one resource should not alter independently of the values of p for other resources.

The most plausible model for evolution is to assume that, for fixed values of the W's and D's, selection will alter the values of p so as to maximise the rate of increase of the predator species. It will also act so as to maximise W/D for those resources for which p is non-zero, selection being weighted in favour of the more abundant resources.

The time which an individual predator spends searching is equal to $1 - \Sigma n_j D_j$. Hence

$$n_j = KR_j p_j (1 - \Sigma n_j D_j), \qquad (72)$$

where K is a constant. This assumes that the number of effective encounters with resource j is proportional to the density of that resource. This in turn is to assume that there is no habitat differentiation such that different resources are abundant in different habitats; it also ignores the formation of a 'search image' by the predator, which could mean that the frequency of effective encounters with different resources was not proportional to their densities. The relevance of habitat differentiation and of search images will be considered later.

The rate of increase of the predator is given by

$$dx/dt = x(\Sigma n_j W_j - T),\qquad(73)$$

where T is a constant, and measures the food required to maintain an individual.

Manipulation of equation (72) gives

$$\Sigma n_j D_j = \frac{K\Sigma p_j R_j D_j}{1 + K\Sigma p_j R_j D_j} = \frac{a}{1+a},\qquad(74)$$

where $a = K\Sigma p_j R_j D_j$.

Using this expression to eliminate n_j from equation (73) gives

$$dx/dt = x\left(\frac{K\Sigma p_j R_j W_j}{1 + K\Sigma p_j R_j D_j} - T\right).\qquad(75)$$

Natural selection will then act on the values of p so as to maximise this expression. Consider first two simple cases.

Case (i) $a \gg 1$. Then $\Sigma n_j D_j \simeq 1$, and the predator spends most of its time pursuing, attacking and consuming its prey, and little time searching.

In this case, selection will maximise

$$\frac{\Sigma p_j R_j W_j}{\Sigma p_j R_j D_j}.$$

This is achieved by putting $p_j = 1$ for that resource for which W_j/D_j is greatest, and $p_j = 0$ for all other resources. That is to say, if resources are abundant but difficult to catch or to consume, selection will lead to specialisation.

The extreme case, $\Sigma n_j D_j = 1$, is unlikely to arise in practice, because it implies that the predator is not resource limited. Thus equation (75) becomes

$$dx/dt = x\left(\frac{W_j}{D_j} - T\right),$$

which implies an unregulated increase or decrease in x with time.

Case (ii) $a \ll 1$. Then $\Sigma n_j D_j \simeq 0$ and the predator spends most of its time searching for food items, which take little time to capture or consume.

In this case, selection will maximise $\Sigma p_j R_j W_j$. This is achieved by putting $p_j = 1$ for all resources for which W_j is positive. That is to say, if resources are difficult to find, selection will produce generalists.

These conclusions are easy to see without the mathematical demonstration. They are clearly stated by MacArthur and Wilson (1967). These authors also consider the effects of competition, and in particular the evolution of 'character divergence', whereby two competing species which overlap in part of their geographical ranges differ more sharply in their characteristics in the zone of overlap. This phenomenon was first noted by Brown and Wilson (1956), who also pointed out that there are two alternative selective mechanisms which might give rise to it, namely selection for species recognition, and selection reducing interspecific competition.

MacArthur and Wilson (1967) analysed further the role of competition in causing character divergence. Unfortunately their analysis is open to criticism. It is based on a model similar to the one I have given, with the additional simplifying assumptions that the resources are 'fine grained', so that they are necessarily taken in the proportions in which they occur in the environment, and that no time is spent in capturing and consuming the resources. The weakness of this method of analysis is that with these assumptions, as has been shown by MacArthur and Wilson themselves, the two predator species could not coexist in the same habitat.

This objection may not be quite fatal to their analysis. Even if two species cannot coexist in a particular habitat in isolation, they may nevertheless be found together in that habitat indefinitely as a result of immigration from elsewhere, as might well occur in a zone of overlap between two species. However, it is worth looking again at the problem of character divergence in terms of the more general model given above.

Competition from a second species will affect the first species by reducing the density of some or all of the resources on which it depends. How will this alter the optimal values of the p_j?

The values of p will be such as to maximise the expression

$$\frac{K\Sigma p_j R_j W_j}{1 + K\Sigma p_j R_j D_j}.$$

Considering a particular p_j, this expression has the form $(e+fp_j)/(g+hp_j)$, where e, f, g, and h are positive constants. It is easy to show that this expression is maximised by putting $p_j = 1$ if $fg - eh > 0$, and otherwise by putting $p_j = 0$. The significance of this inequality is most easily seen by considering just two resources, 1 and 2.

Then the criterion $fg - eh > 0$ becomes

$$\frac{W_2}{D_2} > \frac{KR_1W_1}{1 + KR_1D_1},$$

if it is to be worth eating resource 2 as well as resource 1, and

$$\frac{W_1}{D_1} > \frac{KR_2W_2}{1 + KR_2D_2},$$

if it is to be worth eating resource 1 as well as resource 2.

Note that the significance of these inequalities is that the reduction, through competition, of the density of a resource R_1 or R_2 may make it worth while to take an alternative resource which was not previously worth taking. It will never make it worth while to stop taking a resource which was previously taken. Thus, in the absence of habitat differentiation, two generalist species will not become specialists on different resources in a zone of overlap. But if two species specialising on the same resource have a zone of overlap, one of them may become a generalist in that zone.

This process could contribute to character divergence, but it is unlikely to be the main cause. A more direct effect arises if there is habitat differentiation in a zone of geographical overlap. In such a case, a species which in the absence of competition would occupy both habitats may be confined to one in the zone of overlap.

The conclusions reached in this section can be seen without mathematical demonstration; the purpose of the mathematics is to render the assumptions lying behind the argument more explicit. To summarise briefly, we would expect species which spend most of their time searching for food which, when found, takes little time or effort to capture and consume to be generalists, and species which depend on abundant and easily found resources which take much time and effort to capture or to consume to be specialists. Competition between sympatric species may lead to habitat differences and hence to character divergence. In the absence of habitat differentiation, competition will not convert generalists into specialists, but may cause a specialist species to take food items which it would ignore in the absence of competition.

12 TERRITORIAL BEHAVIOUR

A. The problem

Territorial behaviour, in any animal which displays it, may well be the most important factor stabilising population numbers. Unfortunately, the phenomenon cannot easily be studied in laboratory models. Consequently I will attempt in this chapter to relate the mathematical arguments more directly to field data, and in particular to territorial behaviour in the Great Tit, *Parus major*. However, it is not my intention to give a comprehensive account of the studies which have been made of this bird. Rather, I will use those studies to suggest problems which are worth theoretical analysis.

The difficulty in analysing territorial behaviour lies in the need to consider simultaneously the behavioural mechanisms underlying it, its ecological consequences, and the evolutionary processes which gave rise to it. In particular, it is essential to distinguish between its functions and its consequences. By a function of territorial behaviour I mean a property of that behaviour which has been favoured by natural selection and hence has been responsible for its evolution. If the regulation of population numbers is to be regarded as a function of territorial behaviour, then natural selection must have acted by 'group selection', favouring the survival of some populations at the expense of others. This point was clearly recognised by Wynne-Edwards (1962), who has been the main proponent of the view that the function of territorial behaviour is to regulate populations.

There are serious difficulties in explaining adaptations by group selection (Lack, 1966; Maynard Smith, 1964; Williams, 1966), except in cases where species are divided into small reproductively isolated groups. Lack (1966) and Brown (1969a) have given reasons for thinking that territorial behaviour does not require a group selection explanation. It is therefore assumed here that

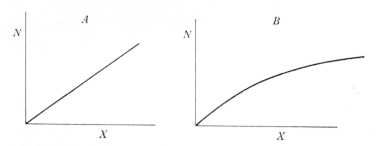

Figure 44. Two possible relationships between N, the number of territories established in an area, and X, the number of pairs attempting to establish territories.

population regulation is not a function of territorial behaviour, and that the latter has evolved because of the increased survival and reproductive success of individuals.

Lack (1968 and earlier) has argued that in many solitary bird species an important function of territory may be to reduce predation. In the case of the Great Tit, Krebs (1970) showed that nests close together were more likely to be predated by weasels. Krebs also argues that territory could function to preserve food supply for the young, and this seems a more generally applicable explanation. Even if boundaries are not defended vigorously after the young hatch out, a pair which ensures that there is no other nest close to their own will thereby guarantee a food supply for the young, without the need for continuing territorial defence.

Whether or not territorial behaviour has consequences for population regulation depends on the relationship between the number of adults in an area immediately prior to the establishment of territories, and the number successfully establishing territories in the area. This relationship could have either of the forms shown in figure 44.

Type A will occur if pairs space themselves out evenly over the available habitat, and if there is no lower limit to the size of territory an individual will accept (or if the lower limit is so small as never to be reached in practice). Lack (1964) has argued that territorial behaviour in the Great Tit follows this pattern. If so, territory has no consequences for population regulation.

Type *B* will occur if there is a lower limit to the acceptable territory size, so that pairs failing to obtain territories of that minimum size leave the area. In this case, territorial behaviour will have consequences for population regulation.

B. A behavioural model

The first step will be to suggest a simple behavioural model of territorial behaviour. The requirements of such a model are that it should ascribe to individuals (or to pairs) a pattern of behaviour which will cause the population to become distributed in space in an appropriate manner. The model will then be used in two ways. First, conclusions will be drawn from it concerning fluctuations in territory size which can be caused by differences in the degree of synchrony of the birds at the start of the breeding season. Second, the model provides a background for the discussion of the evolution of territorial behaviour and of its ecological consequences.

I suppose that during the establishment of territories, each territory has at any moment a 'centre', in the sense that its owner displays with equal energy on either side of this centre, exerting a 'pressure' on his neighbours which declines with l, the distance from the centre. For simplicity I assume that all birds are identical; that is, the pressure they exert is the same function of l. As a result of this display, the boundary between two territories moves so as to equalise the pressure on the two sides; that is, so as to be equidistant from the two centres.

I assume that each individual or pair goes through two phases in its behaviour; different pairs need not be synchronous:

Phase 1. The centre of the territory is not fixed in space. After adjustment of territorial boundaries as explained above, the centre of each territory is adjusted to be midway between boundaries. This is followed by a new adjustment of the boundaries, and the process is continued until a stationary set of boundaries is achieved.

Phase 2. The centre of the territory becomes fixed. This may occur with the selection of a nesting site and the start of nest building. From this point on, only the boundary can be moved.

In either phase, it is assumed that if, when a stationary boundary has been achieved, the distance between two centres is less than $2r$, then one of the two territories will be abandoned, and the remaining territories will be adjusted. This amounts to assuming that there is a minimum acceptable territory size. If there is no minimum distance r, then a type A (figure 44) relation will result.

These are intended to be the simplest behavioural rules which will lead to a region being divided into territories of approximately equal size, with some individuals being excluded from the region. They have some interesting consequences. Consider first a linear habitat (*e.g.* a shore line, a river, a hedgerow). Suppose we have a region of length d, where $d \gg r$. We want to know how the number of established territories depends on the degree of synchrony between different pairs. There are three cases to consider:

(i) Suppose that a number of pairs N, where $N > d/2r$, arrive simultaneously, and that no pairs pass from phase 1 to phase 2 until a stationary state has been reached. Then it is easy to see that the number successfully establishing territories will be $d/2r$, or, more precisely, the nearest integer less than this. The remaining pairs leave the region.

(ii) Suppose now that a number of pairs N, where $N < d/2r$, arrive simultaneously, establish territories and pass to phase 2. The spacing between centres will be $l = d/N$. We ask, what value of N will fill up the region so that no later arrival can establish a territory? It is clear from figure 45 that if $l < 4r$, no newcomer can establish itself. (It is assumed that it is the newcomer rather than an established individual which abandons the region, because a newcomer attempting to establish is attacked with high intensity from both sides.) If $l < 4r$, then $N > d/4r$; that is, any number greater than $d/4r$ will exclude new immigrants. This is half the density which can establish with complete synchrony.

(iii) Suppose finally that each pair arrives, establishes a territory and passes to phase 2 before the arrival of the next. In this case the number finally establishing territories will depend on the positions, presumably arbitrary, in which the first few pairs establish themselves. However, the number will lie between the number attained in case (i) with complete synchrony and half that

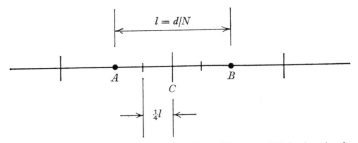

Figure 45. Territories in a linear habitat. Two established pairs have territories with centres at A and B. A newcomer attempts to establish a territory with a centre at C, on the boundary between A and B. Success requires that $\frac{1}{4}l > r$.

number. In small regions the number could fluctuate unpredictably between these densities in different years.

It is not difficult to extend these conclusions to a two-dimensional habitat. With complete synchrony, hexagonal territories of side $2r/\sqrt{3}$ will be established. It is shown in figure 46 that an equally spaced established population at one third of this density will exclude new arrivals.

This model is too artificial to be worth pursuing further. There is, however, some experimental evidence in its favour. Van den Assem (1967) has shown that if male sticklebacks are introduced into a tank simultaneously, nearly twice the density can be attained than if introductions are successive.

An important conclusion from the model is that substantial fluctuations in density could occur because of differences in the degree of synchrony of pairs in different years, without any alteration either in the habitat or in the number of pairs attempting to breed. An important argument in favour of a relationship of type A (figure 44) is that the number of Great Tits breeding in a given wood may fluctuate by a factor of two from year to year. The model suggests an alternative explanation for such fluctuations.

The clearest evidence in favour of a type B relationship in Great Tits is the fact that when established pairs were removed from a stable spring population in a mixed woodland, they were rapidly replaced by new pairs (Krebs, 1971). The newcomers were mainly first-year birds. They came from territories in hedgerows

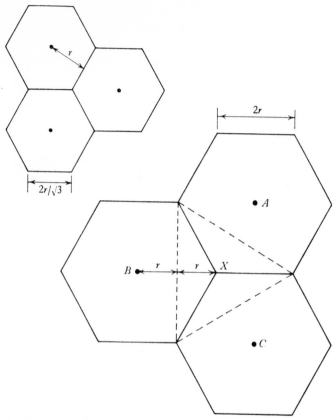

Figure 46. Top left: synchronously established territories with a minimum distance between centres of 2r. Bottom right: established territories with centres A, B and C just permit the establishment of a new territory, centre X, whose boundary is indicated by the broken line.

close to the wood. The vacated hedgerows were not refilled. There is evidence that the hedgerows were suboptimal in terms of reproductive success.

C. Natural selection and territory

It is clear from Krebs' (1971) experiment and from other evidence (e.g. Hinde, 1956) that birds are often excluded from optimal habitats and forced to breed in suboptimal ones. What selective forces are operating to bring this about? Accepting the model of the previous section, there are at least two aspects of behaviour which could be altered by selection:

(i) the minimum acceptable size of territory (r in the model above) in the optimal habitat, and

(ii) the degree of persistence with which an individual attempts to establish a territory in a favourable area rather than move to a less favourable one.

A high degree of persistence would be represented in the model by a large difference between the values of r acceptable in an optimal and a suboptimal habitat. There must be some disadvantages in persistence to counterbalance the advantages of a territory in the optimal habitat. These disadvantages are of two kinds:

(i) those arising from persistence itself – that is, the excessive waste of time and energy in territorial defence, and the risk of injury;

(ii) those arising because a small territory in an optimal habitat may be less productive than a large territory in a suboptimal one, either because of resource limitation or of higher predation.

It is unlikely that the disadvantages are of the first kind only, because if so we would expect the species to evolve a tolerance for a smaller territory. In fact, we would expect selection to adjust r so that the territory size is large enough to contain adequate resources to raise the young, and to adjust the degree of persistence until the reproductive success in the optimal and suboptimal habitats is approximately equal.

One fairly sensitive way of determining whether reproductive success in two habitats is in fact equal was suggested by Brown (1969 b); it is to compare the number of birds reared in the optimal habitat which breed in the suboptimal, and vice versa. Klujver (1951) found that the number of Great Tits reared in deciduous woods (the optimal habitat) and moving to nearby pine woods was exactly the same as the number moving the opposite way. This agrees with the expectation of equal reproductive success.

Krebs (1970), however, found that reproductive success in Great Tits in deciduous woodland was higher than in hedgerows, and that this difference was greater than could be explained by the higher proportion of first year birds in hedgerows. At first sight this result is puzzling. However, the model of territorial behaviour provides one possible explanation. Individuals excluded from the optimal habitat and breeding in the suboptimal one are presumably those which, had they persisted in the optimal habitat, would have been forced to accept a smaller than average territory, either because they were late arrivals or for some other reason. What we expect on theoretical grounds is not equality between the average success in optimal and suboptimal habitats, but equality between the productivity of pairs in the suboptimal habitat and those with the smallest territories in the optimal one. An alternative explanation of Krebs' result is that Great Tits have not yet had time to evolve a perfect adaptation to the environment of modern England.

D. The ecological consequences of territory

It was assumed in the previous section that territorial behaviour has evolved so as to maximise the reproductive success of individuals. Nevertheless it may have consequences for population regulation, which I now consider.

For simplicity, I suppose that two habitats, optimal and suboptimal, are available. I assume that, in both habitats, the reproductive success of a pair depends on the area of the territory they occupy, as shown in figure 47. There are a number of possible patterns of territorial behaviour which could be shown by a species. A classification of these patterns is as follows:

Type 1. 'Spacing territories'

There is no minimum acceptable value of r, and all birds breed in the optimal habitat. This is the only pattern which will give a type A response in figure 44. It corresponds to Fretwell and Lucas' (1969) 'spacing hypothesis'. It is the pattern to be expected if breeding success in the suboptimal habitat is low, or if there is little density dependence of breeding success in the optimal habitat.

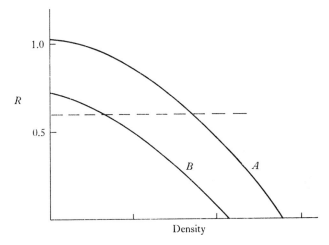

Figure 47. Reproductive success R as a function of density in a favourable (A) and unfavourable (B) habitat. Allowing for adult survival from one season to the next, it is assumed that $R = 0.6$ is sufficient just to maintain the population.

Type 2. 'Fixed minimum territory'

There is a constant minimum value of r, independent of habitat, such that when the optimal habitat is full, breeding success is just sufficient to maintain the population. Pairs establish territories in the optimal habitat if they can. If both habitats are full, additional pairs are excluded from breeding altogether.

With this model, the population will increase until territory size in the suboptimal habitat gives a breeding success sufficient to maintain the population, and so equal to success in the optimal habitat. Hence when the population is at equilibrium, there will be no selection favouring either an increase or a decrease in r.

As the population fluctuates for non-density-dependent reasons, the density in the suboptimal habitat will fluctuate more than in the optimal, so that there may be a 'buffering' effect (Brown, 1969 b), but the average productivity in the two habitats will be equal.

Type 3. *Minimum territory dependent on habitat only*

The minimum acceptable *r* varies with habitat, being smaller in the optimal than the suboptimal habitat. Although not implausible, this model is not investigated further, because of the necessity of introducing a number of additional *ad hoc* assumptions before its consequences can be worked out.

Type 4. *Territory optimising breeding success*

It is supposed that the minimum acceptable *r* is not fixed, but varies both with the habitat and with the densities in the alternative habitats. It is further assumed that each pair finally selects a territory so as to maximise its productivity. There are two possibilities:

Type 4*A*. All territories established synchronously, so that all territories in a given habitat are of the same size, and the productivities in the two habitats are equal. This corresponds to Fretwell and Lucas' (1969) 'density assessment' hypothesis. As in type 2, the average productivities in the two habitats will be equal, but fluctuations in numbers will not be confined to the suboptimal habitat, as expected for type 2.

Type 4*B*. Territories not established synchronously. If, as is likely, the optimal habitat tends to fill up first, then optimising behaviour will equalise the average productivity in the suboptimal habitat with the productivity of the smallest territories in the optimal habitat (that is, what an economist would call the marginal productivity in the optimal habitat). This corresponds to what Fretwell and Lucas call the 'density-limiting' hypothesis – a term I am reluctant to accept, because (as I show in a moment) in all the models except type 1, territorial behaviour is in an important sense density-limiting. In contrast to types 2 and 4*A*, productivity in the suboptimal habitat is consistently lower than in the optimal habitat.

In distinguishing between these patterns of behaviour, type 1 seems to be ruled out if, as in Krebs' (1971) experiment, pairs removed during the breeding season are rapidly replaced. Types 2, 4*A* and 4*B* have the following properties:

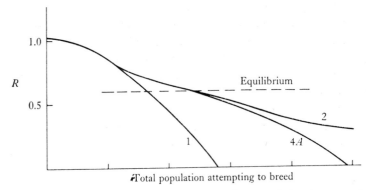

Figure 48. Average reproductive success of a population in an environment consisting of two habitats of equal area, with local reproductive success as shown in figure 47.

1   Birds space themselves evenly in the most favourable habitat only.
2   Birds fill up the optimal habitat first and then the sub-optimal, with a fixed minimum territory size, so that supernumerary birds are excluded from breeding.
4A   No minimum territory size; all birds breed, selecting the habitat which maximises their success.

Type 2.   Fluctuations in numbers mainly in the suboptimal habitat; average breeding success equal.
Type 4A.   Fluctuations in numbers in both habitats; average breeding success equal.
Type 4B.   Fluctuations in numbers in both habitats; average breeding success higher in optimal habitat.

We can now consider the consequences for population dynamics of these patterns of behaviour. The pattern of breeding success shown in figure 47 is assumed throughout. If we also know the relative areas of the two habitats, it is possible to calculate the total productivity of the population as a function of the number of pairs attempting to breed. This has been done in figure 48, assuming equal areas of optimal and suboptimal habitat, for behaviour of types 1, 2 and 4A. The detailed shapes of the curves will depend on the shapes of the curves in figure 47, and on the ratio of the areas, but the qualitative differences do not.

The shallower the slope of R against population size, the more stable the population – that is, the less likely it is to show large amplitude fluctuations. It is clear from figure 48 that although territorial behaviour may increase the reproductive success of the population, it will also stabilise population numbers. The stabilising effect is greatest for territories with a fixed minimum acceptable size. An alternative way of stating the reason for the stabilising effect of territorial behaviour is to say that it acts immediately, by limiting the numbers of breeding pairs, whereas in its absence regulation would be delayed, acting on the survival of the offspring.

REFERENCES

Allee, W. C., Emerson, A. E., Park, O., Park, T. & Schmidt, K. P. 1949. *Principles of Animal Ecology*. Philadephia: W. B. Saunders.

Antonovics, J. & Ford, H. 1972. Criteria for the validation or invalidation of the competitive exclusion principle. *Nature* **237**, 406–8.

Assem, J. van den, 1967. Territory in the three-spined stickleback (*Gasterosteus aculeatus*). *Behaviour*, Suppl. 16, 1–164.

Ayala, F. J. 1969. Experimental invalidation of the principle of competitive exclusion. *Nature* **224**, 1076–9.

Bartlett, M. S. 1966. *An Introduction to Stochastic Processes* (2nd ed.). London: Cambridge University Press.

Brown, J. L. 1969a. Territorial behaviour and population regulation in birds. A review and re-evaluation. *Wilson Bull.* **81**, 293–329.

Brown, J. L. 1969b. The buffer effect and productivity in tit populations. *Amer. Natur.* **103**, 347–54.

Brown, W. L. & Wilson, E. O. 1956. Character displacement. *Syst. Zool.* **5**, 49–64.

Caswell, H. 1972. A simulation study of a time-lag population model. *J. theor. Biol.* **34**, 419–39.

Connell, J. H. 1970. On the role of natural enemies in preventing competitive exclusion in some marine animals and in rain forest trees. In *Proc. Adv. Study Inst. Dynamics Popul. (Oosterbeck)* (eds. P. J. den Boer and G. R. Gradwell), pp. 298–312.

Elton, C. S. 1958. *The Ecology of Invasion by Animals and Plants*. London: Methuen.

Fisher, R. A., Corbet, A. S. & Williams, C. B. 1943. The relation between the number of species and the number of individuals in a random sample of an animal population. *J. Anim. Ecol.* **12**, 42–58.

Fretwell, S. D. & Lucas, H. L. 1969. On territorial behaviour and other factors influencing habitat distribution in birds. I. Theoretical development. *Acta Biotheoretica* **19**, 16–36.

Gardner, M. R. & Ashby, W. R. 1970. Connections of large dynamical (cybernetic) systems: critical values for stability. *Nature* **228**, 784.

Gause, G. F. 1934. *The Struggle for Existence*. Baltimore: Williams and Wilkins.

Gilpin, M. E. & Justice, K. E. 1972. Reinterpretation of the invalidation of the principle of competitive exclusion. *Nature* **236**, 273–301.

Haigh, J. & Maynard Smith, J. 1972. Can there be more predators than prey? *Theoretical Population Biology* **3**, 290–9.

Hardin, G. 1960. The competitive exclusion principle. *Science* **131**, 1292–7.

Hassell, M. P. 1970. Parasite behaviour as a factor contributing to the stability of insect host parasite interactions. In *Proc. Adv. Study Inst. Dynamics Popul. (Oosterbeck)* (eds. P. J. den Boer and G. R. Gradwell), pp. 366–79.

Hassell, M. P. & Varley, G. C. 1969. New inductive population model for insect parasites and its bearing on biological control. *Nature* **223**, 1133–7.

Hinde, R. A. 1956. The biological significance of territories in birds. *Ibis* **98**, 340–69.

Holling, C. S. 1965. The functional response of predators to prey density and its role in mimicry and population regulation. *Mem. Ent. Soc. Can.* **45**, 1–60.

Horne, M. T. 1970. Coevolution of *Escherichia coli* and bacteriophages in chemostat culture. *Science* **168**, 992–3.

Huffaker, C. B. 1958. Experimental studies on predation: dispersion factors and predator–prey oscillations. *Hilgardia* **27**, 343–83.

Hutchinson, G. E. 1948. Circular causal systems in ecology. *Ann. N.Y. Acad. Sci.* **50**, 221–46.

Hutchinson, G. E. 1959. Homage to Santa Rosalia, or why are there so many kinds of animals? *Amer. Natur.* **93**, 145–59.

Kerner, E. H. 1957. A statistical mechanics of interacting biological species. *Bull. Math. Biophys.* **19**, 121–46.

Kerner, E. H. 1959. Further considerations on the statistical mechanics of biological associations. *Bull. Math. Biophys.* **21**, 217–55.

Kimura, M. 1964. *Diffusion Models in Population Genetics.* London: Methuen.

Klujver, H. N. 1951. The population ecology of the Great Tit, *Parus major* L. *Ardea* **39**, 1–135.

Krebs, J. R. 1970. Regulation of numbers in the Great Tit. *J. Zool. Lond.* **162**, 317–33.

Krebs, J. R. 1971. Territory and breeding density in the Great Tit *Parus major* L. *Ecology* **52**, 2–22.

Lack, D. 1947. *Darwin's Finches.* London: Cambridge University Press.

Lack, D. 1954. The evolution of reproductive rates. In *Evolution as a Process* (ed. J. Huxley). London: Allen and Unwin.

Lack, D. 1964. A long term study of the Great Tit (*Parus major*). *J. Anim. Ecol.* **33** (Suppl.), 159–73.

Lack, D. 1966. *Population Studies of Birds.* London: Oxford University Press.

Lack, D. 1968. *Ecological Adaptations for Breeding in Birds.* London: Methuen.

Leigh, E. 1965. On the relation between the productivity, biomass, diversity and stability of a community. *Proc. Natl. Acad. Sci. U.S.* **53**, 777–83.

Leigh, E. 1968. The ecological role of Volterra's equations. In *Some Mathematical Problems of Biology* (ed. M. Gerstenhaber). Providence: The American Mathematical Society.

Leslie, P. H. 1945. The use of matrices in certain population mathematics. *Biometrika* **33**, 183–212.

Leslie, P. H. 1948. Some further notes on the use of matrices in population mathematics. *Biometrika* **35**, 213–45.

Levin, S. A. 1970. Community equilibria and stability, an extension of the competitive exclusion principle. *Amer. Natur.* **104**, 413–23.

Levins, R. 1968. *Evolution in Changing Environments.* Princeton University Press.

Levins, R. 1969. Some demographic and genetic consequences of environmental heterogeneity for biological control. *Bull. Entomol. Soc. Am.* **15**, 237–40.

Levins, R. 1970. Extinction. In *Some Mathematical Problems in Biology* (ed. M. Gerstenhaber). Providence: The American Mathematical Society.

Lewontin, R. C. 1961. Evolution and the theory of games. *J. Theoret. Biol.* **1**, 382–403.

Lewontin, R. C. 1969. The Meaning of Stability. In *Diversity and Stability in Ecological Systems*. Brookhaven Symposium in Biology No. 22, pp. 13–24.

Lewontin, R. C. & Cohen, D. 1969. On population growth in a randomly varying environment. *Proc. Natl. Acad. Sci. U.S.* **62**, 1056–60.

Lotka, A. J. 1925. *Elements of Physical Biology*. Baltimore: Williams and Wilkins.

Luckinbill, L. S. 1973. Coexistence in laboratory populations of *Paramecium aurelia* and its predator *Didinium nasutum*. (In the press.)

MacArthur, R. H. & Levins, R. 1964. Competition, habitat selection and character displacement in a patchy environment. *Proc. Natl. Acad. Sci. U.S.* **51**, 1207–10.

MacArthur, R. H. & Wilson, E. O. 1967. *The Theory of Island Biogeography*. Princeton University Press.

May, R. M. 1971*a*. Stability in model ecosystems. *Proc. Ecol. Soc. Aust.* **6**, 18–56.

May, R. M. 1971*b*. Stability in multispecies community models. *Bull. Math. Biophys.* **12**, 59–79.

May, R. M. 1972. Will a large complex system be stable? *Nature* **238**, 413–14.

May, R. M. & MacArthur, R. H. 1972. Niche overlap as a function of environmental variability. *Proc. Natl. Acad. Sci. U.S.* **69**, 1109–13.

Maynard Smith, J. 1964. Group selection and kin selection. *Nature* **201**, 1145–7.

Maynard Smith, J. 1966. Sympatric speciation. *Amer. Natur.* **100**, 637–50.

Maynard Smith, J. & Slatkin, M. 1973. The stability of predator–prey systems. *Ecology* **54**, 384–91.

Murton, R. K., Westwood, N. J. & Isaacson, A. J. 1964. A preliminary investigation of the factors regulating population size in the Woodpigeon *Columba palumbus*. *Ibis* **106**, 482–507.

Nicholson, A. J. 1954. An outline of the dynamics of animal populations. *Aust. J. Zool.* **2**, 9–65.

Nicholson, A. J. & Bailey, V. A. 1935. The balance of animal populations. Part 1. *Proc. Zool. Soc. Lond.* **3**, 551–98.

Pielou, E. C. 1969. *An Introduction to Mathematical Ecology*. New York: Wiley.

Pimentel, D. 1968. Population regulation and genetic feedback. *Science* **159**, 1432–7.

Pimentel, D. & Al-Hafidh, R. 1965. Ecological control of a parasite population by genetic evolution in a parasite–host system. *Ann. ent. Soc. Am.* **58**, 1–6.

Pimentel, D., Feinberg, E. H., Wood, R. W. & Hayes, J. T. 1965. Selection, spatial distribution and the coexistence of competing fly species. *Amer. Natur.* **99**, 97–109.

Pimentel, D. & Soans, A. B. 1970. Animal populations regulated to carrying capacity of plant host by genetic feedback. In *Proc. Adv. Study Inst. Dynamics Popul. (Oosterbeck)* (eds. P. J. den Boer and G. R. Gradwell), pp. 313–26.

Preston, F. W. 1962. The canonical distribution of commonness and rarity: Part 1. *Ecology* **43**, 185–215; Part 2. *ibid.* **43**, 410–32.

Rescigno, A. & Richardson, I. W. 1965. On the competitive exclusion principle. *Bull. Math. Biophys.* **27**, 85–9.

Rosenzweig, M. L. 1969. Why the prey curve has a hump. *Amer. Natur.* **103**, 81–7.

Rosenzweig, M. L. & MacArthur, R. H. 1963. Graphical representation and stability conditions of predator–prey interactions. *Amer. Natur.* **97**, 209–23.

Satchell, J. E. 1962. Resistance in Oak (*Quercus* spp.) to defoliation by *Tortrix viridana* L. in Roudsea Wood National Nature Reserve. *Ann. appl. Biol.* **50**, 431–42.

Southern, H. N. 1959. Mortality and population control. *Ibis* **101**, 429–36.

Stent, G. S. 1963. *Molecular Biology of Bacterial Viruses*. Folkestone: W. H. Freeman.

Strobeck, C. 1973. N species competition. *Ecology* **54**, 650–4.

Utida, S. 1957. Population fluctuation, an experimental and theoretical approach. *C.S.H. Symp. Quant. Biol.* **22**, 139–51.

Volterra, V. 1926. Variazione e fluttuazini del numero d'individui in specie animali conviventi. *Mem. Accad. Nazionale Lincei* (ser. 6) **2**, 31–113.

Wangersky, P. J. & Cunningham, W. J. 1957. Time lag in prey–predator population models. *Ecology* **38**, 136–9.

Watt, K. E. F. 1964. Comments on fluctuations of animal populations and measures of community stability. *Canad. Ent.* **96**, 1434–42.

Watt, K. E. F. 1965. Community stability and the strategy of biological control. *Canad. Ent.* **97**, 887–95.

Williams, G. C. 1966. *Adaptation and Natural Selection*. Princeton University Press.

Williamson, M. H. 1957. An elementary theory of interspecific competition. *Nature* **180**, 422–5.

Williamson, M. H. 1972. *The Analysis of Biological Populations*. London: Arnold.

Wilson, E. O. & Bossert, W. H. 1971. *A Primer of Population Biology*. Stanford: Sinauer Associates.

Winfree, A. T. 1967. Biological rhythms and the behaviour of populations of coupled oscillators. *J. Theoret. Biol.* **16**, 15–42.

Wynne-Edwards, V. C. 1962. *Animal Dispersion in relation to Social Behaviour*. Edinburgh: Oliver and Boyd.

Zwolfer, H. 1963. The structure of the parasite complexes of some lepidoptera. *Z. angew. Ent.* **51**, 346–57.

INDEX

age distribution of population, 17–18; matrix representation of, 42–3; of predators, 53–6

Al-Hafidh, R., 118

Allee, W. C., 18

Antonovics, J., 62

Ashby, W. R., 107

Assem, J. van den, 129

Ayala, F. J., 62, 64

Bailey, V. A., 49, 50, 52

Bartlett, M. S., 13

biological models, as bridge between mathematical models and ecosystems, 3

biomass, in equations, 92, 95, 109

Bossert, W. H., 85

breeding seasons, discrete: cause delayed regulation of ecosystems, 37, 47–8, 109

Brown, J. L., 125, 131, 133

Brown, W. L., 122

Callosobruchus chinensis, parasitisation of, 51–2

Canadian Forest Insect Survey, 112

carrying capacity (equilibrium density of prey species in absence of predator), 19, 28, 31, 57, 82

Caswell, H., 48

chemical substances: distributions of, as relevant variables in theoretical ecology, 3

climate, destabilising effect of fluctuations in, 85

coevolution, 88–9, 116–24

coexistence of species: conditions for, 59–68, 98; factors favouring, in migration, 79; number of resources and, 86–7, 98–103; selection for ability for, 89

Cohen, D., 14

Columba palumbus, hierarchical behaviour of, 58

commensal interactions between species, 5

competitive exclusion, principle of, 61, 103

competitive interactions between species, 5, 98; analysis of, based on logistic equation, 59–62; with continuous reproduction, 62–4; do not produce oscillations, 59, 67–8; niche overlap and environmental variability in, 65–7; and stability of system, 10, 88, 112

complexity of ecosystem, *see under* stability

computer simulations, 2, 77–9, 107, 110

connectedness (connectance, connectivity) of food web, 95, 96–7, 107; systematic, and stability, 107–12

Connell, J. H., 114–15

conservative dynamical system, 11, 22–3, 25

'conserved quantity' in statistical mechanics, 86, 92, 95

'contest' for resources, tends to stability, 41–2, 56

cover for prey, and predator–prey interaction, 25–6, 30, 56–7, 79

crustacea, filter-feeding on algae, 31

Cunningham, W. J., 48

densities: equilibrium, 106; of parasitoids, search area decreases with increase of, 50, 51; of predator and prey, and rate at which prey are eaten, 19, 24; of prey, response of predator to, 31–2; territorial behaviour and, 126–8, 134

deterministic models, 12–15

development time of an organism, causes delayed regulation of ecosystem, 36–7, 38–42, 48